POST-OFFICE DEPARTMENT.

TABLES OF DISTANCES

FOR THE

REGULATION AND ADJUSTMENT

OF

TELEGRAPHIC RATES FOR GOVERNMENT MESSAGES.

COMPILED BY THE

TOPOGRAPHER OF THE POST-OFFICE DEPARTMENT.

APPROVED BY THE POSTMASTER-GENERAL.

WASHINGTON:
GOVERNMENT PRINTING OFFICE.
1873.

The following tables of distances by which to compute telegraphic rates for Government messages are published for the information of all concerned.

This (first) edition contains a list of more than 2,000 places, comprising all the larger and more important points in the United States.

Their distances by the shortest post-routes are stated from Washington, District of Columbia, and from New York, New York, in two adjoining, parallel columns.

A second list is given of all the military posts of the United States Army, with distances from other points most immediately connected therewith.

This list has been furnished by the Quartermaster-General of the United States Army.

Both tables are of course subject to modifications from time to time, owing to the changing nature of the postal and telegraphic service, and will be extended or otherwise altered as the needs of the service require.

It being impracticable to give the distances by the shortest lines of telegraph, owing to the want of data from the telegraph companies, these distances are given by the shortest post-routes, and are thus available for the computations of mileage in the settlement of accounts for personal travel and other transportation.

JNO. A. J. CRESWELL,
Postmaster-General.

POST-OFFICE DEPARTMENT, *August* 16, 1873.

RATES OF PAY

FOR

COMMUNICATIONS BY TELEGRAPH, 1873-1874.

POST-OFFICE DEPARTMENT,
June 28, 1873.

Whereas, by the act of Congress approved July 24, 1866, entitled "An act to aid in the construction of telegraph lines, and to secure to the Government the use of the same for postal, military, and other purposes," in section second it is enacted, "That telegraphic communications between the several Departments of the Government of the United States and their officers and agents shall, in their transmission over the lines of said companies, have priority over all other business, and shall be sent at rates to be annually fixed by the Postmaster-General:"

Now, therefore, in pursuance and by virtue of the authority on me by said act conferred, I, J. A. J. Creswell, Postmaster-General of the United States, do hereby fix the rates at which the telegraphic communications aforesaid shall be sent, for the year commencing on the first day of July, A. D. eighteen hundred and seventy-three, as follows, namely:

The rate for all telegraphic communications known as the signal-service messages and reports shall be three cents for each word of said reports and messages for each circuit over which it may pass in accordance with the schedule of circuits and plans of the Chief Signal-Officer of the Army, which are now adopted, or may hereafter be adopted by him for transmitting these dispatches, or such part thereof as he may designate, in such words or ciphers as may from time to time be directed by him. The amount thus estimated is to be taken in full payment for said dispatches; no additional allowance to be made for drops, office messages, or other services or special facilities required by the Chief Signal-Officer for the correct and prompt transmission of said signal-service messages and reports.

The rate for all telegraphic communications sent otherwise than over circuits established as aforesaid shall be as follows, viz: One cent per word for each circuit through which it shall be transmitted, said rate to be computed subject to the following conditions, viz:

A distance of two hundred and fifty miles, as computed by the tables of the Post-Office Department, shall be deemed a circuit.

If, in computing circuits, there shall be found one or more circuits and a fraction of a circuit, such fraction shall be deemed a circuit.

If a communication shall be sent a distance less than two hundred and fifty miles, that distance shall be deemed a circuit.

All words of the communication transmitted are to be counted, excepting the date and place at which such communication is filed; all messages of less than twenty-five words, address and signature included, shall be rated as if containing twenty-five words, and all messages exceeding twenty-five words shall be rated by the exact number of words they contain, address and signature included.

JNO. A. J. CRESWELL,
Postmaster-General.

LIST

OF THE

LARGER AND MORE IMPORTANT PLACES

IN

THE UNITED STATES,

WITH

THEIR DISTANCES BY THE SHORTEST POST-ROUTES

TO

WASHINGTON, D. C., AND TO NEW YORK, N. Y.

TABLE OF DISTANCES.

Place, County, and State.	Distance from Washington.	Distance from New York.
	Miles.	*Miles.*
Abbeville C. H., Abbeville, S. C	610	838
Aberdeen, Monroe, Miss	941	1,169
Abingdon, Knox, Ill	919	1,057
Abingdon, Washington, Va	369	597
Abington, Plymouth, Mass	479	251
Ackley, Hardin, Iowa	1,094	1,223
Ada, Hardin, Ohio	549	678
Adams, Berkshire, Mass	400	172
Adams, Jefferson, N. Y	492	314
Adams' Run, Colleton, S. C	603	831
Addison, Steuben, N. Y	325	301
Adrian, Lenawee, Mich	591	720
Afton, Union, Iowa	1,163	1,288
Aiken, Barnwell, S. C	609	837
Akron, Summit, Ohio	423	548
Albany, Dougherty, Ga	823	1,052
ALBANY, Albany, N. Y	373	145
Albany, Linn, Oreg	3,209	3,337
Albert Lea, Freeborn, Minn	1,181	1,310
Albia, Monroe, Iowa	1,083	1,208
Albion, Calhoun, Mich	657	786
Albion, Orleans, N. Y	425	402
Albuquerque, Bernalillo, N. Mex	2,191	2,335
Aledo, Mercer, Ill	950	1,079
Alexandria, Alexandria, Va	8	236
Algona, Kossuth, Iowa	1,172	1,301
Allegan, Allegan, Mich	733	862
Allegheny, Allegheny, Pa	304	433
Allentown, Lehigh, Pa	191	92
Alliance, Stark, Ohio	387	516
Alpena, Alpena, Mich	855	931
Alton, Madison, Ill	918	1,055
Altona, Knox, Ill	920	1,049
Altoona, Blair, Pa	254	314
Amboy, Lee, Ill	872	1,001
Amenia, Dutchess, N. Y	316	88
Americus, Sumter, Ga	788	1,016
Ames, Story, Iowa	1,098	1,227
Amesbury, Essex, Mass	506	278
Amherst, Hampshire, Mass	402	159
Amsterdam, Montgomery, N. Y	406	178
Anamosa, Jones, Iowa	982	1,111
Anderson, Madison, Ind	656	767
Anderson C. H., Anderson, S. C	633	861
Andover, Essex, Mass	487	259
Angelica, Allegany, N. Y	398	374
Anna, Union, Ill	924	1,104

Place, County, and State.	Distance from Washington.	Distance from New York.
	Miles.	*Miles.*
ANNAPOLIS, Anne Arundel, Md	42	226
Annapolis Junction, Anne Arundel, Md	22	206
Ann Arbor, Washtenaw, Mich	629	716
Anoka, Anoka, Minn	1,215	1,344
Ansonia, New Haven, Conn	302	74
Apalachicola, Franklin, Fla	1,074	1,302
Appleton, Outagamie, Wis	985	1,114
Arcola, Douglas, Ill	801	938
Arlington, Middlesex, Mass	474	246
Arlington, Phelps, Mo	1,033	1,177
Ashburnham, Worcester, Mass	459	231
Asheville, Buncombe, N. C	496	724
Ashland, Ashland, Ohio	473	602
Ashland, Schuylkill, Pa	209	190
Ashland, Hanover, Va	99	327
Ashley, Washington, Ill	863	1,043
Ashtabula, Ashtabula, Ohio	431	547
Astoria, Clatsop, Oreg	3,250	3,378
Atchison, Atchison, Kans	1,210	1,335
Athens, Clarke, Ga	707	935
Athens, Greene, N. Y	345	117
Athens, Bradford, Pa	319	259
Athens, Athens, Ohio	404	619
Athol, Worcester, Mass	418	190
Athol Depot, Worcester, Mass	417	189
ATLANTA, Fulton, Ga	725	953
Atlanta, Logan, Ill	860	997
Atlantic, Cass, Iowa	1,208	1,287
Atlantic City, Atlantic, N. J	199	126
Atlantic City, Sweet Water, Wyo	2,228	2,357
Atoka, Choctaw Nation, Ind. T	1,464	1,608
Attica, Fountain, Ind	754	881
Attica, Wyoming, N. Y	415	391
Attleborough, Bristol, Mass	433	205
Auburn, Placer, Cal	3,005	3,134
Auburn, Androscoggin, Me	607	379
Auburn, Cayuga, N. Y	414	318
Augusta, Richmond, Ga	592	820
AUGUSTA, Kennebec, Me	636	408
Aurora, Kane, Ill	811	940
Aurora, Dearborn, Ind	588	768
Aurora, Esmeralda, Nev	3,037	3,166
Aurora, Cayuga, N. Y	397	299
Austin, Mower, Minn	1,113	1,242
Austin, Lander, Nev	2,741	2,870
AUSTIN, Travis, Tex	1,706	1,934
Avoca, Steuben, N. Y	342	318
Avon, Livingston, N. Y	390	366
Babylon, Suffolk, N. Y	265	37
Bagdad, Shelby, Ky	724	904

Place, County, and State.	Distance from Washington.	Distance from New York.
	Miles.	*Miles.*
Bainbridge, Decatur, Ga	918	1,146
Bainbridge, Chenango, N. Y	390	366
Bald Mountain, Gilpin, Colo	1,871	2,015
Baldwin, Duval, Fla	925	1,153
Baldwinsville, Onondaga, N. Y	448	305
Ballston, Saratoga, N. Y	404	177
Baltimore, Baltimore, Md	38	190
Bangor, Penobscot, Me	710	482
Bannack City, Beaver Head, Mont	2,669	2,798
Baraboo, Sauk, Wis	941	1,070
Bardstown, Nelson, Ky	712	892
Bardstown Junction, Bullett, Ky	694	874
Barnesville, Belmont, Ohio	376	528
Barnstable, Barnstable, Mass	471	243
Barre, Worcester, Mass	410	182
Bastrop, Bastrop, Tex	1,742	1,930
Batavia, Kane, Ill	820	949
Batavia, Genesee, N. Y	426	402
Batesville, Independence, Ark	1,186	1,414
Batesville, Panola, Miss	996	1,224
Bath, Sagadahoc, Me	608	380
Bath, Steuben, N. Y	334	310
Baton Rouge, East Baton Rouge, La	1,318	1,546
Battle Creek, Calhoun, Mich	682	799
Baxter Springs, Cherokee, Kans	1,261	1,405
Bayard, Columbiana, Ohio	384	513
Bay City, Bay, Mich	713	789
Beardstown, Cass, Ill	916	1,053
Beatrice, Gage, Nebr	1,369	1,494
Beaufort, Carteret, N. C	378	606
Beaufort, Beaufort, S. C	654	882
Beaver, Beaver, Pa	331	460
Beaver Dam, Dodge, Wis	922	1,051
Beaver Falls, Beaver, Pa	333	462
Becket, Berkshire, Mass	412	184
Bedford, Lawrence, Ind	701	881
Belchertown, Hampshire, Mass	392	164
Belfast, Waldo, Me	704	476
Belle Air, Harford, Md	71	178
Bellaire, Belmont, Ohio	349	501
Bellefield, Greencastle, Va	181	409
Bellefontaine, Logan, Ohio	552	686
Bellefonte, Centre, Pa	266	296
Belle Plaine, Benton, Iowa	1,027	1,156
Belleville, Saint Clair, Ill	920	1,064
Belleville, Richland, Ohio	493	622
Bellevue, Huron, Ohio	511	640
Bellows Falls, Windham, Vt	452	224
Beloit, Rock, Wis	863	992
Belvidere, Boone, Ill	850	979
Belvidere, Warren, N. J	235	78
Bement, Piatt, Ill	812	949
Benicia, Solano, Cal	3,105	3,229
Bennington, Bennington, Vt	411	183
Benton Harbor, Berrien, Mich	774	903

Place, County, and State.	Distance from Washington.	Distance from New York.
	Miles.	*Miles.*
Bent's Fort, Bent, Colo	1,746	1,890
Berea, Cuyahoga, Ohio	457	586
Berkeley Springs, Morgan, W. Va	105	321
Berlin, Green Lake, Wis	955	1,084
Berrien Springs, Berrien, Mich	730	850
Berwick, Columbia, Pa	217	242
Bethel, Oxford, Me	642	414
Bethlehem, Northampton, Pa	196	87
Beverly, Essex, Mass	482	254
Beverly, Burlington, N. J	154	78
Biddeford, York, Me	557	329
Big Rapids, Mecosta, Mich	786	916
Binghamton, Broome, N. Y	356	215
Black Hawk Point, Gilpin, Colo	1,867	2,011
Black River Falls, Jackson, Wis	1,037	1,166
Blackstone, Worcester, Mass	423	195
Blairstown, Benton, Iowa	1,017	1,146
Blairsville, Indiana, Pa	321	381
Blairsville Junction, Indiana, Pa	318	378
Bloomfield, Davis, Iowa	1,068	1,193
Bloomfield, Essex, N. J	224	16
Bloomington, McLean, Ill	840	977
Bloomington, Monroe, Ind	727	871
Bloomsburgh, Columbia, Pa	201	203
Blossburgh, Tioga, Pa	257	283
Blue Earth City, Faribault, Minn	1,229	1,358
Bluffton, Wells, Ind	646	775
BOISÉ CITY, Ada, Idaho	2,639	2,769
Bonham, Fannin, Tex	1,469	1,697
Boonesborough, Boone, Iowa	1,114	1,243
Booneville, Cooper, Mo	1,093	1,237
Boonton, Morris, N. J	258	32
Boonville, Oneida, N. Y	503	275
Bordentown, Burlington, N. J	167	64
Boscobel, Grant, Wis	974	1,103
BOSTON, Suffolk, Mass	464	236
Boston, Bowie, Tex	1,268	1,496
Bowling Green, Warren, Ky	786	966
Bozeman, Gallatin, Mont	2,705	2,834
Bradford, Miami, Ohio	570	707
Bradford, Orange, Vt	520	292
Brandon, Rankin, Miss	1,002	1,230
Brandon, Rutland, Vt	482	254
Branford, New Haven, Conn	312	84
Brashear, Saint Mary's, La	1,270	1,498
Brattleborough, Windham, Vt	428	200
Brazil, Clay, Ind	731	868
Brazos, Santiago, Cameron, Tex	1,735	1,963
Breckinridge, Wilkin, Minn	1,405	1,534
Bremond, Robertson, Tex	1,539	1,768
Brenham, Washington, Tex	1,608	1,794
Brewster's Station, Putnam, N. Y	283	55
Bricksburgh, Ocean, N. J	200	64
Bridgeport, Fairfield, Conn	287	59
Bridgeport, Belmont, Ohio	353	505

Place, County, and State.	Distance from Washington.	Distance from New York.
	Miles.	*Miles.*
Bridgeton, Cumberland, Me	617	389
Bridgeton, Cumberland, N. J	179	128
Bridgewater, Plymouth, Mass	439	211
Brighton, Macoupin, Ill	931	1,068
Brighton, Middlesex, Mass	462	234
Brimfield, Peoria, Ill	896	1,034
Bristol, Hartford, Conn	335	107
Bristol, Grafton, N. H	537	309
Bristol, Bucks, Pa	158	66
Bristol, Bristol, R. I	435	207
Bristol, Sullivan, Tenn	384	612
Brockport, Monroe, N. Y	413	389
Brocton, Chautauqua, N. Y	491	467
Brookfield, Worcester, Mass	398	170
Brodhead, Greene, Wis	882	1,011
Brookfield, Linn, Mo	1,089	1,226
Brookhaven, Lincoln, Miss	1,071	1,299
Brookline, Norfolk, Mass	470	240
Brooklyn, Poweshiek, Iowa	1,059	1,188
Brooklyn, Kings, N. Y	230	2
Brookville, Franklin, Ind	608	788
Brookville, Saline, Kans	1,391	1,535
Brookville, Jefferson, Pa	329	389
Brownsville, Fayette, Pa	272	487
Brownsville, Haywood, Tenn	941	1,169
Brownsville, Cameron, Tex	1,770	1,998
Brownville, Nehama, Nebr	1,261	1,398
Brunswick, Glynn, Ga	779	1,007
Brunswick, Cumberland, Me	599	371
Brunswick, Chariton, Mo	1,091	1,235
Bryan, Williams, Ohio	612	741
Bryan, Brazos, Tex	1,538	1,765
Bryan, Uintah, Wyo	2,124	2,253
Buchanan, Berrien, Mich	731	860
Buchanan, Allegheny, Pa	305	434
Bucksport, Hancock, Me	722	494
Bucyrus, Crawford, Ohio	504	633
Buffalo, Erie, N. Y	446	422
Bunker Hill, Macoupin, Ill	899	1,036
Burlingame, Osage, Kans	1,285	1,429
Burlington, Des Monies, Iowa	983	1,108
Burlington, Burlington, N. J	158	69
Burlington, Chittenden, Vt	531	303
Burlington, Racine, Wis	857	986
Burnham Village, Waldo, Me	669	441
Burr Oak, Saint Joseph, Mich	663	792
Bushnell, McDonough, Ill	932	1,069
Butler, Bates, Mo	1,174	1,318
Butler, Butler, Pa	377	437

Place, County, and State.	Distance from Washington.	Distance from New York.
	Miles.	*Miles.*
Cadiz, Harrison, Ohio	379	508
Cairo, Alexander, Ill	913	1,085
Calais, Washington, Me	833	605
Calera, Shelby, Ala	799	1,027
California, Moniteau, Mo	1,056	1,200
Calmar, Winneshiek, Iowa	1,045	1,174
Calumet, Houghton, Mich	1,280	1,411
Calvert, Robertson, Tex	1,527	1,755
Cambria, Columbia, Wis	939	1,068
Cambridge, Henry, Ill	918	1,056
Cambridge, Middlesex, Mass	468	240
Cambridge, Dorchester, Md	226	236
Cambridge, Washington, N. Y	409	181
Cambridge, Guernsey, Ohio	401	553
Cambridge City, Wayne, Ind	622	759
Cambridgeport, Middlesex, Mass	470	242
Camden, Ouachita, Ark	1,172	1,400
Camden, Knox, Me	720	492
Camden, Camden, N. J	141	89
Camden, Oneida, N. Y	492	273
Camden, Kershaw, S. C	557	785
Cameron, Clinton, Mo	1,156	1,293
Camp Apache, Maricopa, Ariz	2,480	2,624
Camp Bidwell, Siskiyou, Cal	3,140	3,269
Camp Bowie, Pima, Ariz	2,785	2,937
Camp Brown, Wyo	2,236	2,375
Camp Crittenden, Pima, Ariz	2,724	2,868
Camp Date Creek, Yavapai, Ariz	3,210	3,339
Camp Douglas, Salt Lake, Utah	2,338	2,467
Camp Gaston, Klamath, Cal	3,311	3,440
Camp Grant, Pima, Ariz	2,733	2,877
Camp Halleck, Elko, Nev	2,603	2,732
Camp Harney, Grant, Oreg	2,982	3,111
Camp Hualpai, Yavapai, Ariz	3,331	3,460
Camp Lowell, Pima, Ariz	2,673	2,817
Camp McDowell, Maricopa, Ariz	2,853	2,942
Camp Mohave, Mohave, Ariz	2,930	3,059
Camp Stockton, Presidio, Tex	2,162	2,390
Camp Three Forks, Owyhee, Idaho	2,745	2,873
Camp Verde, Yavapai, Ariz	3,194	3,323
Camp Warner, Grant, Oreg	3,489	3,319
Camp Wright, Mendocino, Cal	3,318	3,447
Canajoharie, Montgomery, N. Y	428	201
Canandaigua, Ontario, N. Y	366	342
Canastota, Madison, N. Y	456	272
Candor, Tioga, N. Y	344	246
Cannelton, Perry, Ind	799	979
Cannonsburgh, Washington, Pa	325	454
Canton, Fulton, Ill	898	1,035
Canton, Norfolk, Mass	449	221
Canton, Madison, Miss	1,040	1,268
Canton, Lewis, Mo	1,002	1,139
Canton, Saint Lawrence, N. Y	565	386
Canton, Stark, Ohio	405	534
Canton, Van Zandt, Tex	1,685	1,827
Canyon City, Grant, Oreg	2,819	2,948

Place, County, and State.	Distance from Washington.	Distance from New York.
	Miles.	*Miles.*
Cape Disappointment, Wash	3,264	3,393
Cape Girardeau, Cape Girardeau, Mo	1,027	1,214
Cape May, Cape May, N. J	209	157
Cape Vincent, Jefferson, N. Y	570	351
Carbondale, Jackson, Ill	904	1,076
Carbondale, Luzerne, Pa	276	163
Cardington, Morrow, Ohio	514	643
Carey, Wyandot, Ohio	545	674
Carlisle, Cumberland, Pa	141	201
Carlyle, Clinton, Ill	857	1,029
Carmansville, New York, N. Y	235	7
Carrollton, Greene, Ill	939	1,076
Carrollton, Carroll, Ky	623	803
Carrollton, Carroll, Mo	1,115	1,259
CARSON CITY, Ormsby, Nev	2,921	3,050
Cartersville, Bartow, Ga	678	906
Carthage, Hancock, Ill	962	1,100
Carthage, Jasper, Mo	1,234	1,378
Carthage, Jefferson, N. Y	563	314
Carver, Carver, Minn	1,212	1,341
Cascade, Dubuque, Iowa	998	1,127
Cassopolis, Cass, Mich	726	855
Castile, Wyoming, N. Y	389	365
Castleton, Rutland, Vt	454	226
Catasauqua, Lehigh, Pa	194	97
Catlettsburgh, Boyd, Ky	456	684
Catskill, Greene, N. Y	341	113
Cazenovia, Madison, N. Y	471	187
Cedar Falls, Black Hawk, Iowa	1,061	1,190
Cedar Keys, Levy, Fla	1,049	1,277
Cedar Rapids, Linn, Iowa	992	1,121
Centerville, Appanoose, Iowa	1,117	1,246
Central City, Gilpin, Colo	869	2,013
Central Falls, Providence, R. I	425	197
Centralia, Marion, Ill	849	1,021
Centreville, Queen Anne, Md	145	155
Centreville, Saint Joseph, Mich	701	830
Ceredo, Wayne, W. Va	451	679
Cerro Gordo, Piatt, Ill	821	958
Chagrin Falls, Cuyahoga, Ohio	434	563
Chambersburgh, Franklin, Pa	99	235
Champaign, Champaign, Ill	792	929
Champlain, Clinton, N. Y	591	363
Chardon, Geauga, Ohio	485	570
Chariton, Lucas, Iowa	1,113	1,238
Charles City, Floyd, Iowa	1,092	1,221
Charleston, Coles, Ill	791	928
Charleston, Charleston, S. C	576	804
Charleston, Bradley, Tenn	584	812
Charleston, Kanawha, W. Va	330	558
Charlestown, Clarke, Ind	677	857
Charlestown, Sullivan, N. H	460	232
Charlestown, Jefferson, W. Va	65	280
Charlestown, Middlesex, Mass	465	237
Charlotte, Eaton, Mich	672	801
Charlotte, Mecklenburgh, N. C	398	625

Place, County, and State.	Distance from Washington.	Distance from New York.
	Miles.	*Miles.*
Charlottesville, Albemarle, Va	118	346
Chateaugay, Franklin, N. Y	632	404
Chatfield, Fillmore, Minn	1,125	1,254
Chatham Four Corners, Columbia, N. Y	358	130
Chatsworth, Livingston, Ill	787	925
Chattanooga, Hamilton, Tenn	626	854
Chelsea, Suffolk, Mass	468	240
Chelsea, Orange, Vt	525	297
Chenoa, McLean, Ill	809	947
Cheraw, Chesterfield, S. C	513	741
Cherrystone, Northampton, Va	270	344
Cherry Valley, Otsego, N. Y	441	213
Cheshire, New Haven, Conn	319	91
Chester, Randolph, Ill	917	1,089
Chester, Delaware, Pa	123	103
Chester, Chesterfield, Va	128	356
Chester C. H., Chester, S. C	443	671
Chestertown, Kent, Md	169	179
Chetopah, Labette, Kans	1,278	1,422
Cheyenne Agency, Dak	1,672	1,801
Cheyenne City, Laramie, Wyo	1,782	1,911
Cheyenne Wells, Greenwood, Colo	1,653	1,797
Chicago, Cook, Ill	772	901
Chico, Butte, Cal	3,100	3,229
Chicopee, Hampden, Mass	371	143
Chicopee Falls, Hampden, Mass	374	146
Chillicothe, Peoria, Ill	875	1,013
Chillicothe, Ross, Ohio	465	680
Chillicothe, Livingston, Mo	1,115	1,252
Chippewa Falls, Chippewa, Wis	1,111	1,240
Chittenango, Madison, N. Y	450	279
Christiansburgh, Montgomery, Va	265	493
Cincinnati, Hamilton, Ohio	564	744
Circleville, Pickaway, Ohio	468	643
Claremont, Sullivan, N. H	469	241
Clarence, Cedar, Iowa	958	1,087
Clarence, Shelby, Mo	1,044	1,181
Clarinda, Page, Iowa	1,224	1,349
Clarion, Clarion, Pa	391	520
Clarksburgh, Harrison, W. Va	276	491
Clarksville, Johnson, Ark	1,176	1,404
Clarksville, Pike, Mo	985	1,122
Clarksville, Montgomery, Tenn	840	1,068
Clarksville, Red River, Tex	1,421	1,649
Clayton, Kent, Del	146	156
Clearfield, Clearfield, Pa	281	341
Cleaveland, Bradley, Tenn	597	825
Cleveland, Cuyahoga, Ohio	444	573
Clifton Springs, Ontario, N. Y	377	356
Clinton, Clinton, Iowa	911	1,040
Clinton, DeWitt, Ill	834	971
Clinton, East Feliciana, La	1,165	1,393
Clinton, Worcester, Mass	438	210
Clinton, Henry, Mo	1,135	1,279
Clinton, Oneida, N. Y	487	249
Clinton, Rock, Wis	850	979

Place, County, and State.	Distance from Washington.	Distance from New York.
	Miles.	*Miles.*
Clover, Halifax, Va	210	438
Cloverdale, Sonoma, Cal	3,206	3,335
Clyde, Wayne, N. Y	438	331
Clyde, Sandusky, Ohio	519	648
Coatesville, Chester, Pa	142	128
Cobleskill, Schoharie, N. Y	453	190
Cohoes, Albany, N. Y	383	155
Colchester, New London, Conn	378	150
Cold Spring, Putnam, N. Y	282	54
Coldwater, Branch, Mich	645	774
Collinsville, Hartford, Conn	343	115
Colorado City, El Paso, Colo	1,905	2,049
Colorado Springs, El Paso, Colo	1,904	2,048
Columbia, Boone, Mo	1,050	1,194
Columbia, Lancaster, Pa	109	175
COLUMBIA, Richland, S. C	507	735
Columbia, Maury, Tenn	823	1,051
Columbia City, Whitley, Ind	642	770
Columbus, Muscogee, Ga	816	1,044
Columbus, Bartholomew, Ind	687	824
Columbus, Hickman, Ky	961	1,136
Columbus, Lowndes, Miss	968	1,195
Columbus, Platte, Nebr	1,358	1,487
COLUMBUS, Franklin, Ohio	487	624
Columbus, Colorado, Tex	1,610	1,838
Columbus, Columbia, Wis	922	1,056
Colusa, Colusa, Cal	3,081	3,210
Company's Shop, Alamance, N. C	327	555
Concord, Middlesex, Mass	478	250
Concord, Cabarrus, N. C	376	604
CONCORD, Merrimack, N. H	505	277
Conneaut, Ashtabula, Ohio	444	519
Conneautville, Crawford, Pa	416	527
Connellsville, Fayette, Pa	246	461
Connersville, Fayette, Ind	630	771
Conshohocken, Montgomery, Pa	152	100
Constantine, Saint Joseph, Mich	685	814
Cooperstown, Otsego, N. Y	439	236
Cooperstown Junction, Otsego, N. Y	423	220
Corinne, Box Elder, Utah	2,323	2,452
Corinth, Alcorn, Miss	844	1,072
Corning, Steuben, N. Y	314	291
Corpus Christi, Nueces, Tex	1,753	1,979
Corry, Erie, Pa	429	455
Corsicana, Navarro, Tex	1,654	1,796
Cortland Village, Cortland, N. Y	400	258
Corunna, Shiawassee, Mich	677	753
Corvallis, Benton, Oreg	3,219	3,347
Corydon, Wayne, Iowa	1,151	1,280
Coshocton, Coshocton, Ohio	427	556
Cottonwood Springs, Lincoln, Nebr	1,555	1,684
Council Bluffs, Pottawattomie, Iowa	1,262	1,391
Council Grove, Morris, Kans	1,342	1,486
Covington, Fountain, Ind	746	883
Covington, Kenton, Ky	566	746

2

Place, County, and State.	Distance from Washington.	Distance from New York.
	Miles.	*Miles.*
Covington, Alleghany, Va	226	454
Coxsackie, Greene, N. Y	352	123
Crawfordsville, Montgomery, Ind	718	855
Cresco, Howard, Iowa	1,064	1,193
Cresson, Cambria, Pa	270	330
Crestline, Crawford, Ohio	492	621
Crow Creek Agency, Dak	1,521	1,650
Crown Point, Lake, Ind	760	899
Cuba, Allegany, N. Y	406	382
Culpeper, Culpeper, Va	70	298
Cumberland, Alleghany, Md	153	368
Cuthbert, Randolph, Ga	835	1,063
Cuyahoga Falls, Summit, Ohio	426	555
Cynthiana, Harrison, Ky	632	812
Cypress City, Harris, Tex	1,601	1,829
Dallas, Dallas, Tex	1,600	1,742
Dalton, Whitfield, Ga	625	853
Danbury, Fairfield, Conn	297	69
Danielsonville, Windham, Conn	393	165
Dansville, Livingston, N. Y	366	342
Danvers, Essex, Mass	485	257
Danville, Vermilion, Ill	760	897
Danville, Hendricks, Ind	693	830
Danville, Boyle, Ky	702	882
Danville, Montour, Pa	192	215
Danville, Pittsylvania, Va	256	484
Dardanelle, Yell, Ark	1,183	1,411
Darien, McIntosh, Ga	756	984
Darlington, La Fayette, Wis	940	1,069
Darlington C. H., Darlington, S. C	483	711
Davenport, Scott, Iowa	956	1,085
Dayton, Montgomery, Ohio	558	695
Dearbornville, Wayne, Mich	617	688
Decatur, Morgan, Ala	749	977
Decatur, Macon, Ill	833	970
Decatur, Van Buren, Mich	729	858
Decherd, Franklin, Tenn	695	923
Decorah, Winneshiek, Iowa	1,057	1,186
Dedham, Norfolk, Mass	448	220
Defiance, Defiance, Ohio	608	737
DeKalb Centre, DeKalb, Ill	830	959
Delanco, Burlington, N. J.	152	75
Delavan, Tazewell, Ill	871	1,008
Delavan, Walworth, Wis	863	992
Delaware, Delaware, Ohio	511	648
Delaware City, New Castle, Del	132	142
Delhi, Delaware, N. Y	396	202
Delphi, Carroll, Ind	705	843
Delphos, Van Wert, Ohio	578	707

Place, County, and State.	Distance from Washington.	Distance from New York.
	Miles.	*Miles.*
Demopolis, Marengo, Ala	903	1,131
DENVER, Arapahoe, Colo	1,829	1,973
De Pere, Brown, Wis	1,008	1,137
Deposit, Broome, N. Y	394	136
Derby, New Haven, Conn	301	72
Derby Line, Orleans, Vt	601	373
DES MOINES, Polk, Iowa	1,129	1,258
De Soto, Dallas, Iowa	1,151	1,280
Detroit, Wayne, Mich	623	678
Devall's Bluff, Prairie, Ark	1,024	1,252
DeWitt, Clinton, Iowa	930	1,059
Dexter, Penobscot, Me	697	469
Dexter, Washtenaw, Mich	638	725
Dixon, Lee, Ill	870	999
Dodgeville, Iowa, Wis	958	1,087
Donaldsonville, Ascension, La	1,268	1,496
DOVER, Kent, Del	157	167
Dover, Strafford, N. H	532	304
Dover, Morris, N. J	249	43
Dowagiac, Cass, Mich	740	869
Downieville, Sierra, Cal	3,054	3,183
Downingtown, Chester, Pa	173	122
Doylestown, Bucks, Pa	172	120
Dresden, Muskingum, Ohio	443	572
Driftwood, Cameron, Pa	299	324
Drum Barracks, Los Angeles, Cal	3,317	3,446
Dryden, Tompkins, N. Y	366	268
Dublin, Wayne, Ind	624	761
Dubuque, Dubuque, Iowa	962	1,091
Du Luth, Saint Louis, Minn	1,344	1,483
Dundee, Yates, N. Y	333	309
Dunkirk, Chautauqua, N. Y	483	459
Duquoin, Perry, Ill	884	1,056
Durham's, Orange, N. C	326	554
Dwight, Livingston, Ill	838	976
Eagle Harbor, Keweenaw, Mich	1,277	1,408
Eagle Pass, Maverick, Tex	1,965	2,193
Eagle River, Keweenaw, Mich	1,268	1,399
Earlville, La Salle, Ill	846	975
East Cambridge, Middlesex, Mass	467	239
East Greenwich, Kent, R. I	402	174
East Hampton, Hampshire, Mass	375	147
East Liverpool, Columbiana, Ohio	347	481
East New York, Kings, N. Y	234	6
Easton, Talbot, Md	190	200
Easton, Northampton, Pa	206	76
Eastport, Washington, Me	829	601
East Saginaw, Saginaw, Mich	700	776

Place, County, and State.	Distance from Washington.	Distance from New York.
	Miles.	*Miles.*
East Saint Louis, Saint Clair, Ill	905	1,049
Eastville, Northampton, Va	328	338
Eaton, Preble, Ohio	617	753
Eaton Rapids, Eaton, Mich	661	790
Eatonton, Putnam, Ga	706	934
Eau Claire, Eau Claire, Wis	1,093	1,222
Ebensburgh, Cambria, Pa	282	342
Eddyville, Wapello, Iowa	1,075	1,200
Edenton, Chowan, N. C	299	469
Edgartown, Dukes, Mass	481	253
Edinburgh, Johnson, Ind	668	834
Edwardsville, Madison, Ill	910	1,047
Effingham, Effingham, Ill	814	951
Egg Harbor, Atlantic, N. J	182	130
Eldora, Hardin, Iowa	1,089	1,218
Elgin, Kane, Ill	814	943
Elizabeth, Union, N. J	212	15
Elizabethport, Union, N. J	213	16
Elizabethtown, Hardin, Ky	714	894
Elkader, Clayton, Iowa	1,025	1,154
Elkhart, Elkhart, Ind	690	819
Elkhorn, Walworth, Wis	868	997
Elko, Elko, Nev	2,573	2,702
Elkton, Cecil, Md	91	137
Ellenville, Ulster, N. Y	338	112
Ellicott City, Howard, Md	37	205
Ellicottsville, Cattaraugus, N. Y	448	424
Ellsworth, Ellsworth, Kans	1,414	1,558
Ellsworth, Hancock, Me	735	507
Elmira, Chemung, N. Y	297	273
El Paso, Woodford, Ill	824	962
El Paso, El Paso, Tex	2,460	2,688
Elyria, Lorain, Ohio	470	599
Elyton, Jefferson, Ala	768	996
Eminence, Henry, Ky	658	838
Emmittsburgh, Frederick, Md	102	254
Emporia, Lyon, Kans	1,319	1,463
Emporium, Cameron, Pa	318	343
Englewood, Bergen, N. J	241	15
Enterprise, Clark, Miss	936	1,164
Erie, Erie, Pa	467	492
Esconawba, Delta, Mich	1,084	1,216
Essex, Middlesex, Conn	339	111
Eufaula, Barbour, Ala	860	1,088
Eugene City, Lane, Oreg	3,252	3,380
Eureka, Humboldt, Cal	3,408	3,537
Eureka, Woodford, Ill	838	976
Eutaw, Greene, Ala	859	1,087
Evanston, Cook, Ill	784	913
Evansville, Vanderburgh, Ind	809	979
Evansville, Rock, Wis	888	1,017
Exeter, Rockingham, N. H	514	286

Place, County, and State.	Distance from Washington.	Distance from New York.
	Miles.	*Miles.*
Fairbury, Livingston, Ill	798	936
Fairfield, Jefferson, Iowa	1,033	1,158
Fair Haven, New Haven, Conn	306	78
Fairhaven, Bristol, Mass	445	217
Fair Haven, Rutland, Vt	458	230
Fairmount, Marion, W. Va	276	492
Fairport, Monroe, N. Y	405	363
Fall River, Bristol, Mass	418	190
Falmouth, Pendleton, Ky	605	785
Fargo, Grand Forks, Dak	1,549	1,678
Faribault, Rice, Minn	1,162	1,291
Farmington, Fulton, Ill	893	1,031
Farmington, Franklin, Me	653	425
Farmland, Randolph, Ind	608	745
Farmville, Prince Edward, Va	185	413
Fayetteville, Washington, Ark	1,340	1,568
Fayetteville, Cumberland, N. C	382	610
Fayetteville, Onondaga, N. Y	447	304
Fayetteville, Lincoln, Tenn	735	963
Fentonville, Genesee, Mich	653	728
Fernandez de Taos, Taos, N. Mex	1,977	2,121
Fernandina, Nassau, Fla	972	1,200
Fillmore City, Millard, Utah	2,485	2,614
Findley, Hancock, Ohio	560	689
Fisherville, Merrimack, N. H	512	284
Fishkill on the Hudson, Dutchess, N. Y	288	60
Fitchburgh, Worcester, Mass	448	220
Flemingsburgh, Fleming, Ky	681	861
Flemington, Hunterdon, N. J	198	52
Flint, Genesee, Mich	666	742
Flora, Clay, Ill	809	981
Florence, Lauderdale, Ala	798	1,026
Florence, Darlington, S. C	473	701
Flushing, Queens, N. Y	244	16
Fonda, Montgomery, N. Y	417	189
Fond du Lac, Fond du Lac, Wis	949	1,078
Fordham, Westchester, N. Y	240	12
Forest, Hardin, Ohio	533	662
Foreston, Ogle, Ill	892	1,021
Fort Abercrombie, Shyenne, Dak	1,419	1,548
Fort Arbuckle, Chickasaw N., Ind. T	1,520	1,664
Fort Atkinson, Jefferson, Wis	883	1,012
Fort Barrancas, Escambia, Fla	1,033	1,261
Fort Bascom, San Miguel, N. Mex	2,157	2,303
Fort Bayard, Grant, N. Mex	2,535	2,679
Fort Belknap, Young, Tex	1,676	1,834
Fort Benton, Choteau, Mont	2,912	3,041
Fort Bliss, El Paso, Tex	2,463	2,691
Fort Boisé, Ada, Idaho	2,639	2,769
Fort Brady, Chippewa, Mich	962	1,017
Fort Bridger, Uintah, Wyo	2,178	2,307
Fort Buford, Howard, Dak	2,059	2,188
Fort Clark, Kinney, Tex	1,889	2,117
Fort Columbus, New York, N. Y	229	1
Fort Colville, Stevens, Wash	3,163	3,292

Place, County, and State	Distance from Washington.	Distance from New York.
	Miles.	*Miles.*
Fort Concho, Bexar, Tex	1,859	2,003
Fort Craig, Socorro, N. Mex	2,311	2,455
Fort Cummings, Grant, N. Mex	2,490	2,634
Fort D. A. Russell, Laramie, Wyo	1,784	1,913
Fort Davis, Presidio, Tex	2,150	2,294
Fort Delaware, New Castle, Del	132	143
Fort Dodge, Webster, Iowa	1,235	1,283
Fort Dodge, Ford, Kans	1,558	1,702
Fort Duncan, Maverick, Tex	1,950	2,178
Fort Edward, Washington, N. Y	428	200
Fort Ellis, Gallatin, Mont	2,778	2,907
Fort Fred Steele, Carbon, Wyo	1,963	2,092
Fort Garland, Costilla, Colo	1,921	2,065
Fort Gibson, Cherokee N., Ind. T	1,370	1,514
Fort Gratiot, Saint Clair, Mich	686	619
Fort Griffin, Shackelford, Tex	1,719	1,863
Fort Hamilton, Kings, N. Y	236	8
Fort Harker, Ellsworth, Kans	1,409	1,553
Fort Hays, Ellis, Kans	1,479	1,623
Fort Howard, Brown, Wis	972	998
Fort Independence, Suffolk, Mass	466	238
Fort Jackson, Plaquemines, La	1,262	1,490
Fort Jefferson, Monroe, Fla	1,443	1,354
Fort Kearney, Kearney, Nebr	1,462	1,591
Fort Klamath, Wasco, Oreg	3,279	3,408
Fort Knox, Waldo, Me	726	498
Fort La Fayette, Kings, N. Y	236	8
Fort Laramie, Laramie, Wyo	1,862	1,991
Fort Larned, Pawnee, Kans	1,525	1,669
Fort Leavenworth, Leavenworth, Kans	1,218	1,362
Fort Lyon, Bent, Colo	1,728	1,872
Fort Mackinac, Mackinac, Mich	952	1,007
Fort McIntosh, Webb, Tex	1,971	2,199
Fort McKavett, Menard, Tex	1,914	2,058
Fort McPherson, Lincoln, Nebr	1,549	1,678
Fort McRae, Socorro, N. Mex	2,341	2,485
Fort Madison, Lee, Iowa	1,002	1,127
Fort Monroe, Elizabeth City, Va	223	375
Fort Niagara, Niagara, N. Y	482	458
Fort Pickering, Essex, Mass	480	252
Fort Plain, Montgomery, N. Y	431	203
Fort Popham, Cumberland, Me	625	397
Fort Quitman, El Paso, Tex	2,249	2,385
Fort Randall, Todd, Dak	1,424	1,553
Fort Ransom, Ransom, Dak	1,492	1,621
Fort Reynolds, Pueblo, Colo	1,806	1,950
Fort Rice, Morton, Dak	1,864	1,993
Fort Richardson, Jack, Tex	1,600	1,828
Fort Ripley, Morrison, Minn	1,313	1,442
Fort Sanders, Albany, Wyo	1,836	1,965
Fort Schuyler, Westchester, N. Y	245	17
Fort Scott, Bourbon, Kans	1,206	1,350
Fort Sedgwick, Weld, Colo	1,646	1,775
Fort Selden, Doña Ana, N. Mex	2,409	2,553
Fort Sewall, Essex, Mass	484	256

Place, County, and State.	Distance from Washington.	Distance from New York.
	Miles.	*Miles.*
Fort Shaw, Lewis and Clarke, Mont	2,852	2,977
Fort Sill, Choctaw N., Ind. T	1,626	1,770
Fort Smith, Sebastian, Ark	1,241	1,506
Fort Snelling, Hennepin, Minn	1,195	1,324
Fort Stanton, Socorro, N. Mex	2,359	2,503
Fort Stevens, Clatsop, Oreg	3,231	3,360
Fort Stevenson, Stevens, Dak	1,992	2,121
Fort Sully, Buffalo, Dak	1,609	1,738
Fort Sumner, San Miguel, N. Mex	2,300	2,444
Fort Totten, Ramsey, Dak	1,592	1,721
Fort Union, Mora, N. Mex	1,846	1,990
Fort Wallace, Wallace, Kans	1,613	1,757
Fort Warren, Suffolk, Mass	471	243
Fort Wayne, Allen, Ind	622	751
Fort Yuma, San Diego, Cal	3,027	3,171
Forsyth, Monroe, Ga	742	970
Fostoria, Seneca, Ohio	552	681
Foxborough, Norfolk, Mass	442	214
Fox Lake, Dodge, Wis	928	1,057
Framingham, Middlesex, Mass	447	119
FRANKFORT, Franklin, Ky	694	874
Franklin, Johnson, Ind	671	831
Franklin, Simpson, Ky	830	986
Franklin, Saint Mary's, La	1,308	1,536
Franklin, Norfolk, Mass	432	204
Franklin, Merrimack, N. H	524	296
Franklin, Venango, Pa	426	509
Franklin, Williamson, Tenn	800	1,028
Franklin, Warren, Ohio	574	711
Franklin, Pendleton, W. Va	190	418
Franklin Grove, Lee, Ill	860	989
Franklinville, Randolph, N. C	344	572
Frederick, Frederick, Md	57	251
Frederick Junction, Frederick, Md	54	248
Fredericksburgh, Spottsylvania, Va	55	283
Fredonia, Chautauqua, N. Y	486	462
Freeburgh, Saint Clair, Ill	927	1,071
Freehold, Monmouth, N. J	199	56
Freeport, Stephenson, Ill	893	1,022
Freeport, Armstrong, Pa	356	416
Fremont, Dodge, Nebr	1,313	1,442
Fremont, Sandusky, Ohio	532	661
Frenchtown, Hunterdon, N. J	202	90
Friendship, Allegany, N. Y	397	373
Frostburgh, Alleghany, Md	171	386
Fulton, Whitesides, Ill	908	1,037
Fulton, Callaway, Mo	1,039	1,183
Fulton, Oswego, N. Y	460	313
Fultonville, Montgomery, N. Y	418	190

Place, County, and State.	Distance from Washington.	Distance from New York.
	Miles.	*Miles.*
Gainesville, Sumter, Ala	1,032	1,260
Gainesville, Alachua, Fla	976	1,204
Galena, Jo Daviess, Ill	943	1,072
Galesburgh, Knox, Ill	910	1,048
Galesburgh, Kalamazoo, Mich	696	825
Galion, Crawford, Ohio	493	622
Gallatin, Gallatin, Mont	2,839	2,968
Gallatin, Sumner, Tenn	805	1,011
Gallipolis, Gallia, Ohio	485	700
Galt, Sacramento, Cal	3,068	3,197
Galva, Henry, Ill	913	1,042
Galveston, Galveston, Tex	1,490	1,718
Gardiner, Kennebec, Me	628	400
Gardner, Worcester, Mass	444	216
Garnett, Anderson, Kans	1,267	1,411
Garrettsville, Portage, Ohio	409	538
Garrison's, Putnam, N. Y	279	51
Geneseo, Henry, Ill	916	1,045
Geneseo, Livingston, N. Y	400	376
Geneva, Kane, Ill	808	937
Geneva, Ontario, N. Y	388	344
Geneva, Ashtabula, Ohio	433	557
Geneva, Walworth, Wis	843	972
Georgetown, Clear Creek, Colo	1,886	2,030
Georgetown, Scott, Ky	682	862
Georgetown, Clay, Minn	1,467	1,596
Georgetown, Georgetown, S. C	555	783
Gettysburgh, Adams, Pa	115	230
Gilman, Iroquois, Ill	771	909
Gilroy, Santa Clara, Cal	3,196	3,325
Girard, Macoupin, Ill	896	1,033
Girard, Crawford, Kans	1,232	1,376
Girard, Erie, Pa	436	522
Glasgow, Barren, Ky	773	953
Glasgow Junction, Barren, Ky	763	943
Glasgow, Howard, Mo	1,086	1,214
Glen Mills, Delaware, Pa	137	112
Glen Rock, York, Pa	80	204
Glen's Falls, Warren, N. Y	434	206
Glenwood, Mills, Iowa	1,254	1,379
Gloucester, Essex, Mass	496	268
Gloversville, Fulton, N. Y	427	199
Golden City, Jefferson, Colo	1,846	1,990
Gold Hill, Storey, Nev	2,910	3,039
Goldsborough, Wayne, N. C	281	509
Gonzales, Gonzales, Tex	1,686	1,914
Gordonsville, Orange, Va	97	325
Gorham, Coos, N. H	663	435
Goshen, Elkhart, Ind	681	810
Goshen, Orange, N. Y	286	60
Gouverneur, Saint Lawrence, N. Y	542	363
Grafton, Worcester, Mass	428	200
Grafton, Taylor, W. Va	380	470
Grand Haven, Ottawa, Mich	763	867
Grand Junction, Hardeman, Tenn	885	1,113

Place, County, and State.	Distance from Washington.	Distance from New York.
	Miles.	*Miles.*
Grand Rapids, Kent, Mich	732	861
Grand Rapids, Wood, Wis	1,104	1,233
Granville, Licking, Ohio	459	598
Grass Valley, Nevada, Cal	3,000	3,129
Great Barrington, Berkshire, Mass	372	144
Great Bend, Susquehanna, Pa	307	194
Great Falls, Strafford, N. H	538	310
Green Bay, Brown, Wis	969	1,101
Greencastle, Putnam, Ind	712	849
Greencastle, Franklin, Pa	88	246
Greene, Chenango, N. Y	368	226
Greeneville, Greene, Tenn	440	668
Greenfield, Hancock, Ind	654	791
Greenfield, Franklin, Mass	403	175
Greenfield, Highland, Ohio	490	705
Green Point, Kings, N. Y	232	4
Greenport, Suffolk, N. Y	325	98
Greensborough, Hale, Ala	911	1,139
Greensborough, Guilford, N. C	305	533
Greensburgh, Decatur, Ind	632	812
Greensburgh, Westmoreland, Pa	340	400
Greenville, Butler, Ala	906	1,134
Greenville, New London, Conn	368	140
Greenville, Bond, Ill	863	1,000
Greenville, Montcalm, Mich	733	822
Greenville, Washington, Miss	1,217	1,445
Greenville, Darke, Ohio	581	718
Greenville, Mercer, Pa	388	517
Greenville C. H., Greenville, S. C.	651	879
Greenville, Hunt, Tex	1,599	1,743
Greenwich, Fairfield, Conn	260	32
Greenwich, Washington, N. Y	410	182
Grenada, Grenada, Miss	985	1,213
Griffin, Spalding, Ga	769	997
Griggsville, Pike, Ill	936	1,073
Grinnell, Poweshiek, Iowa	1,074	1,203
Groton Junction, Middlesex, Mass	450	222
Guilford, New Haven, Conn	320	92
Hackensack, Bergen, N. J	240	14
Hackettstown, Warren, N. J	232	62
Hagerstown, Washington, Md	77	257
Hagerstown Junction, Washington, Md	53	269
Halleck Station, Elko, Nev	2,550	2,679
Hallowell, Kennebec, Me	633	405
Hallsville, Harrison, Tex	1,332	1,560
Hamburgh, Fremont, Iowa	1,263	1,388
Hamilton, Caldwell, Mo	1,141	1,278
Hamilton, White Pine, Nev	2,731	2,860
Hamilton, Madison, N. Y	418	269

Place, County, and State.	Distance from Washington.	Distance from New York.
	Miles.	*Miles.*
Hamilton, Butler, Ohio	589	730
Hammonton, Atlantic, N. J	171	121
Hampton, Elizabeth City, Va	226	378
Hancock, Washington, Md	96	312
Hancock, Houghton, Mich	1,266	1,398
Hannibal, Marion, Mo	973	1,110
Hanover, Grafton, N. H	496	268
Hanover, York, Pa	98	213
Harper's Ferry, Jefferson, W. Va	55	271
Harrington, Kent, Del	173	183
HARRISBURGH, Dauphin, Pa	123	183
Harrison, Hamilton, Ohio	590	770
Harrisonburgh, Rockingham, Va	146	374
Harrisonville, Cass, Mo	1,160	1,304
HARTFORD, Hartford, Conn	341	113
Harvard, McHenry, Ill	835	964
Hastings, Barry, Mich	699	828
Hastings, Dakota, Minn	1,163	1,292
Havana, Mason, Ill	907	1,036
Havana, Schuyler, N. Y	316	292
Haverhill, Essex, Mass	497	269
Haverstraw, Rockland, N. Y	271	43
Havre de Grace, Harford, Md	74	153
Hays City, Ellis, Kans	1,479	1,623
Hazleton, Luzerne, Pa	244	146
Healdsburgh, Sonoma, Cal	3,188	3,317
Helena, Phillips, Ark	1,025	1,253
HELENA, Lewis and Clarke, Mont	2,769	2,897
Helena, Karnes, Tex	1,710	1,938
Hempstead, Queens, N. Y	253	25
Hempstead, Austin, Tex	1,587	1,815
Henderson, Henderson, Ky	821	991
Henry, Marshall, Ill	899	1,028
Herkimer, Herkimer, N. Y	454	226
Hernando, De Soto, Miss	971	1,199
Hickman, Fulton, Ky	971	1,179
Highland, Madison, Ill	882	1,019
Hightstown, Mercer, N. J	180	52
Hillsborough, Montgomery, Ill	869	1,006
Hillsborough, Orange, N. C	340	568
Hillsborough, Highland, Ohio	543	758
Hillsdale, Hillsdale, Mich	624	753
Hingham, Plymouth, Mass	481	253
Hinsdale, Cheshire, N. H	422	194
Hoboken, Hudson, N. J	229	3
Holden, Johnson, Mo	1,138	1,282
Holland, Ottawa, Mich	756	885
Hollidaysburgh, Blair, Pa	262	335
Holliston, Middlesex, Mass	450	221
Holly, Oakland, Mich	649	725
Holly Springs, Marshall, Miss	910	1,138
Holmes's Hole, Dukes, Mass (Vineyard Haven P. O.)	471	243
Holyoke, Hampden, Mass	375	147
Homer, Champaign, Ill	779	916

Place, County, and State.	Distance from Washington.	Distance from New York.
	Miles.	*Miles.*
Homer, Cortland, N. Y	403	261
Homewood, Beaver, Pa	338	467
Honeoye Falls, Monroe, N. Y	385	361
Honesdale, Wayne, Pa	293	135
Hookset, Merrimack, N. Y	494	266
Hoosick Falls, Rensselaer, N. Y	407	179
Hopkinsville, Christian, Ky	852	1,028
Hornellsville, Steuben, N. Y	355	331
Horseheads, Chemung, N. Y	303	279
Hot Springs, Hot Spring, Ark	1,136	1,364
Houghton, Houghton, Mich	1,314	1,398
Houlton, Aroostook, Me	837	609
Houma, Terre Bonne, La	1,259	1,487
Houston, Harris, Tex	1,540	1,768
Howell, Livingston, Mich	645	730
Hudson, Lenawee, Mich	608	737
Hudson, Hudson, N. J	229	3
Hudson, Columbia, N. Y	344	116
Hudson, Summit, Ohio	418	547
Hudson, Saint Croix, Wis	1,161	1,290
Humboldt, Allen, Kans	1,301	1,445
Humboldt, Gibson, Tenn	917	1,145
Hunter's Point, Kings, N. Y	232	4
Huntington, Huntington, Ind	646	775
Huntington, Suffolk, N. Y	266	38
Huntingdon, Huntingdon, Pa	220	280
Huntsville, Madison, Ala	724	952
Huntsville, Walker, Tex	1,514	1,742
Hyannis, Barnstable, Mass	477	249
Hyde Park, Dutchess, N. Y	308	80
Hyde Park, Luzerne, Pa	259	149
Idaho, Clear Creek, Colo	1,867	2,011
Idaho City, Boisé Idaho	2,674	2,803
Ilion, Herkimer, N. Y	456	228
Independence, Buchanan, Iowa	1,031	1,160
Independence, Jackson, Mo	1,178	1,322
Indiana, Indiana, Pa	338	398
INDIANAPOLIS, Marion, Ind	674	811
Indianola, Warren, Iowa	1,150	1,279
Indianola, Calhoun, Tex	1,620	1,846
Iola, Allen, Kans	1,293	1,437
Ionia, Ionia, Mich	712	801
Iowa City, Johnson, Iowa	1,009	1,138
Iowa Falls, Hardin, Iowa	1,105	1,234
Ipswich, Essex, Mass	491	263
Ironton, Iron, Mo	993	1,137
Ironton, Lawrence, Ohio	507	722
Irvington, Essex, N. J	221	13
Irvine, Warren, Pa	407	432

Place, County, and State.	Distance from Washington.	Distance from New York.
	Miles.	*Miles.*
Irwin's Station, Westmoreland, Pa	350	410
Ishpeming, Marquette, Mich	1,149	1,281
Island Pond, Essex, Vt	721	493
Ithaca, Tompkins, N. Y	367	271
Iuka, Tishemingo, Miss	821	1,049
Jacksborough, Jack, Tex	1,650	1,794
Jackson, Amador, Cal	3,107	3,236
Jackson, Jackson, Mich	637	766
JACKSON, Hinds, Miss	1,016	1,244
Jackson, Jackson, Ohio	447	662
Jackson, Madison, Tenn	900	1,128
Jacksonport, Jackson, Ark	1,153	1,381
Jacksonville, Duval, Fla	945	1,172
Jacksonville, Morgan, Ill	904	1,041
Jacksonville, Jackson, Oreg	3,339	3,468
Jamaica, Queens, N. Y	241	14
Jamaica Plain, Norfolk, Mass	459	231
Jamestown, Chautauqua, N. Y	423	447
Janesville, Rock, Wis	863	992
Jefferson, Greene, Iowa	1,143	1,272
Jefferson, Ashtabula, Ohio	420	544
Jefferson, Marion, Tex	1,334	1,562
Jefferson, Jefferson, Wis	889	1,018
Jefferson Barracks, Saint Louis, Mo	916	1,060
JEFFERSON CITY, Cole, Mo	1,031	1,175
Jeffersonville, Clark, Ind	674	854
Jersey City, Hudson, N. J	227	2
Jersey Shore, Lycoming, Pa	231	256
Jerseyville, Jersey, Ill	938	1,075
Johnstown, Fulton, N. Y	423	195
Johnstown, Cambria, Pa	293	353
Joliet, Will, Ill	784	939
Jonesborough, Washington, Tenn	416	644
Jonesville, Hillsdale, Mich	627	756
Jordan, Onondaga, N. Y	453	310
Julesburgh, Weld, Colo	1,643	1,772
Junction City, Davis, Kans	1,329	1,473
Kalamazoo, Kalamazoo, Mich	705	834
Kanawha C. H., (Charleston,) Kanawha, W. Va	330	558
Kankakee, Kankakee, Ill	777	915
Kansas City, Jackson, Mo	1,189	1,333
Katonah, Westchester, N. Y	273	45
Keene, Cheshire, N. H	441	213

Place, County, and State.	Distance from Washington.	Distance from New York.
	Miles.	*Miles.*
Keeseville, Essex, N. Y	532	304
Kelton, Box Elder, Utah	2,389	2,518
Kendall's Mills, Somerset, Me	659	431
Kendallville, Noble, Ind	648	777
Kennebunk, York, Me	549	321
Kennedy, Chautauqua, N. Y	462	438
Kennett's Square, Chester, Pa	114	125
Kenosha, Kenosha, Wis	824	953
Kent, Portage, Ohio	413	542
Kentland, Newton, Ind	743	881
Kenton, Hardin, Ohio	545	674
Keokuk, Lee, Iowa	1,071	1,109
Kewanee, Henry, Ill	904	1,033
Key Port, Monmouth, N. J	213	25
Key West, Monroe, Fla	1,379	1,607
Kilbourn City, Columbia, Wis	967	1,096
Kinderhook, Columbia, N. Y	365	137
Kingston, Bartow, Ga	666	894
Kingston, Ulster, N. Y	322	94
Kingston, Luzerne, Pa	243	164
Kingston, Washington, R. I	389	161
Kingsville, Richland, S. C	532	760
Kirksville, Adair, Mo	1,055	1,192
Kit Carson, Greenwood, Colo	1,678	1,822
Kittaning, Armstrong, Pa	370	430
Kittery, York, Me	522	294
Knightstown, Henry, Ind	641	778
Knobnoster, Johnson, Mo	1,114	1,258
Knoxville, Knox, Ill	905	1,043
Knoxville, Marion, Iowa	1,111	1,236
Knoxville, Knox, Tenn	514	742
Kokomo, Howard, Ind	676	813
Lacon, Marshall, Ill	900	1,029
Laconia, Belknap, N. H	532	304
La Crosse, La Crosse, Wis	1,054	1,183
La Fayette, Tippecanoe, Ind	722	860
La Grange, Troup, Ga	797	1,025
La Grange, La Grange, Ind	668	797
La Grange, Oldham, Ky	646	826
La Grange, Fayette, Tex	1,686	1,914
Lake City, Columbia, Fla	885	1,113
Lake City, Wabashaw, Minn	1,125	1,254
Lake Forest, Lake, Ill	802	931
Lake Village, Belknap, N. H	534	306
Lambertville, Hunterdon, N. J	186	74
Lanark, Carroll, Ill	914	1,043
Lancaster, Garrard, Ky	698	878
Lancaster, Coos, N. H	573	345
Lancaster, Fairfield, Ohio	446	629

Place, County, and State.	Distance from Washington.	Distance from New York.
	Miles.	*Miles.*
Lancaster, Lancaster, Pa	121	158
Lancaster, Grant, Wis	966	1,095
Lansing, Allamakee, Iowa	1,031	1,160
LANSING, Ingham, Mich	674	763
Lansingburgh, Rensselaer, N. Y	382	155
La Paz, Yuma, Ariz	3,135	3,279
Lapeer, Lapeer, Mich	686	664
La Porte, La Porte, Ind	717	846
Laramie City, Albany, Wyo	1,839	1,968
Laredo, Webb, Tex	1,971	2,199
La Salle, La Salle, Ill	871	1,000
Latrobe, Westmoreland, Pa	330	390
Lawrence, Douglas, Kans	1,229	1,373
Lawrence, Essex, Mass	480	252
Lawrenceburgh, Dearborn, Ind	586	766
Lawrenceburgh, Armstrong, Pa	409	469
Lawton, Clinch, Ga	813	1,041
Lawton, Van Buren, Mich	71	850
Leavenworth City, Leavenworth, Kans	1,215	1,359
Lebanon, Saint Clair, Ill	881	1,053
Lebanon, Boone, Ind	702	839
Lebanon, Marion, Ky	739	919
Lebanon, Laclede, Mo	1,081	1,235
Lebanon, Grafton, N. H	495	267
Lebanon, Warren, Ohio	546	719
Lebanon, Lebanon, Pa	149	157
Lebanon, Wilson, Tenn	810	1,038
Lee, Berkshire, Mass	386	158
Leesburgh, Loudoun, Va	46	266
Lee's Summit, Jackson, Mo	1,166	1,310
Leetonia, Columbiana, Ohio	366	495
Lemont, Cook, Ill	797	926
Lena, Stephenson, Ill	898	1,027
Lenox, Berkshire, Mass	389	161
Leominster, Worcester, Mass	447	219
Le Roy, Genesee, N. Y	404	382
Lewes, Sussex, Del	213	223
Lewisburgh, Conway, Ark	1,121	1,349
Lewisburgh, Union, Pa	189	215
Lewisburgh, Greenbrier, W. Va	256	484
Lewiston, Androscoggin, Me	607	379
Lewistown, Fulton, Ill	918	1,055
Lewistown, Mifflin, Pa	186	246
Lexington, Fayette, Ky	665	845
Lexington, Sanilac, Mich	707	640
Lexington, La Fayette, Mo	1,150	1,294
Lexington, Davidson, N. C	337	565
Lexington, Rockbridge, Va	193	421
Liberty, Clay, Mo	1,196	1,333
Liberty, Bedford, Va	203	431
Ligonier, Noble, Ind	665	794
Lima, Livingston, N. Y	389	365
Lima, Allen, Ohio	564	693
Lincoln, Logan, Ill	864	1,001
LINCOLN, Lancaster, Nebr	1,333	1,462

Place, County, and State.	Distance from Washington.	Distance from New York.
	Miles.	*Miles.*
Linwood Station, Delaware, Pa	119	107
Litchfield, Litchfield, Conn	336	108
Litchfield, Montgomery, Ill	880	1,017
Little Falls, Herkimer, N. Y	447	218
LITTLE ROCK, Pulaski, Ark	1,072	1,300
Littleton, Grafton, N. H	551	323
Live Oak, Suwannee, Fla	862	1,090
Livingston, Sumter, Ala	885	1,113
Lockhaven, Clinton, Pa	244	269
Lockport, Will, Ill	804	933
Lockport, Niagara, N. Y	452	428
Lodi, Columbia, Wis	930	1,059
Logan, Hocking, Ohio	430	645
Logansport, Cass, Ind	685	823
London, Madison, Ohio	508	645
Long Branch, Monmouth, N. J	222	32
Longview, Upshur, Tex	1,342	1,570
Los Angeles, Los Angeles, Cal	3,297	3,441
Los Lumas, Valencia, N. Mex	2,213	2,357
Loudonville, Ashland, Ohio	460	589
Louisiana, Pike, Mo	965	1,102
Louisville, Jefferson, Ky	672	852
Lowell, Middlesex, Mass	467	239
Lowell, Kent, Mich	727	816
Lowville, Lewis, N. Y	509	298
Ludlow, Windsor, Vt	479	251
Lumberton, Robeson, N. C	433	661
Lynchburgh, Campbell, Va	179	407
Lynn, Essex, Mass	475	247
Lyons, Clinton, Iowa	913	1,042
Lyons, Wayne, N. Y	431	338
McConnellsville, Morgan, Ohio	454	612
McGregor, Clayton, Iowa	1,004	1,133
Machias, Washington, Me	792	564
McKeesport, Allegheny, Pa	288	447
McKenzie, Carroll, Tenn	899	1,116
McMinnville, Warren, Tenn	742	970
McPherson, Lincoln, Nebr	1,543	1,672
Mackinaw, Mackinac, Mich	925	858
Macomb, McDonough, Ill	959	1,067
Macon, Bibb, Ga	717	945
Macon, Noxubee, Miss	975	1,203
Macon City, Macon, Mo	1,055	1,192
Madison, Morgan, Ga	696	924
Madison, Jefferson, Ind	661	841
Madison, Carroll, N. H	570	342
Madison, Morris, N. J	235	27
MADISON, Dane, Wis	904	1,033
Magnolia, Duplin, N. C	317	545

Place, County, and State.	Distance from Washington.	Distance from New York.
	Miles.	*Miles.*
Mahanoy City, Schuylkill, Pa	212	186
Malden, Middlesex, Mass	469	241
Malone, Franklin, N. Y	618	416
Manassas, Prince William, Va	35	263
Manchester, Delaware, Iowa	1,009	1,138
Manchester, Hillsborough, N. H	487	259
Manchester, Bennington, Vt	436	208
Manhattan, Riley, Kans	1,309	1,453
Manistee, Manistee, Mich	776	958
Manitowoc, Manitowoc, Wis	936	1,065
Mankato, Blue Earth, Minn	1,191	1,320
Mansfield, Richland, Ohio	479	608
Maquoketa, Jackson, Iowa	949	1,078
Marblehead, Essex, Mass	484	255
Mare Island, Solano, Cal	3,100	3,229
Marengo, McHenry, Ill	838	967
Marengo, Iowa, Iowa	1,039	1,168
Maricopa Wells, Pima, Ariz	2,827	2,971
Marietta, Cobb, Ga	704	932
Marietta, Washington, Ohio	370	585
Marietta, Lancaster, Pa	113	179
Marion, Perry, Ala	890	1,118
Marion, Grant, Ind	644	781
Marion, Linn, Iowa	998	1,127
Marion, Wayne, N. Y	423	358
Marion, Marion, Ohio	513	642
Marion, Smyth, Va	339	567
Marlborough, Middlesex, Mass	452	224
Marlin, Falls, Tex	1,558	1,786
Maroa, Macon, Ill	845	982
Marquette, Marquette, Mich	1,160	1,291
Marseilles, La Salle, Ill	848	977
Marshall, Clark, Ill	764	901
Marshall, Calhoun, Mich	669	798
Marshall, Harrison, Tex	1,318	1,546
Marshalltown, Marshall, Iowa	1,062	1,191
Marshfield, Webster, Mo	1,123	1,267
Martinsburgh, Berkeley, W. Va	74	289
Martinsville, Morgan, Ind	698	841
Maysville, Yuba, Cal	3,056	3,185
Marysville, Union, Ohio	520	657
Maryville, Nodaway, Mo	1,232	1,357
Mason, Effingham, Ill	836	963
Mason, Ingham, Mich	662	791
Mason City, Mason, Ill	885	1,022
Mason City, Cerro Gordo, Iowa	1,119	1,248
Massillon, Stark, Ohio	413	542
Matawan, Monmouth, N. J	210	67
Matteawan, Dutchess, N. Y	289	61
Mattoon, Coles, Ill	802	939
Mauch Chunk, Carbon, Pa	220	122
Maumee City, Lucas, Ohio	567	696
Mauston, Juneau, Wis	986	1,115
Maysville, Mason, Ky	695	875
Meadville, Crawford, Pa	454	496

Place, County, and State.	Distance from Washington.	Distance from New York.
	Miles.	*Miles.*
Mechanicsburgh, Cumberland, Pa	132	192
Mechanicsville, Bucks, Pa	177	124
Medford, Middlesex, Mass	470	242
Media, Delaware, Pa	133	103
Medina, Orleans, N. Y	435	411
Medina, Medina, Ohio	448	578
Melrose, Middlesex, Mass	471	243
Memphis, Shelby, Tenn	937	1,165
Menasha, Winnebago, Wis	958	1,087
Mendota, La Salle, Ill	856	985
Mendota, Dakota, Minn	1,208	1,337
Menomonee, Menomonee, Mich	1,066	1,195
Menomonee, Dunn, Wis	1,116	1,245
Mercer, Mercer, Pa	405	534
Merchantville, Steuben, N. Y	327	303
Meredith Village, Belknap, N. H	542	314
Meriden, New Haven, Conn	322	94
Meridian, Lauderdale, Miss	921	1,149
Mesilla, Doña Aña, N. Mex	2,397	2,541
Methuen, Essex, Mass	482	254
Metropolis City, Massac, Ill	909	1,089
Mexico, Audrian, Mo	1,014	1,158
Mexico, Oswego, N. Y	488	310
Michigan City, La Porte, Ind	730	859
Middleborough, Plymouth, Mass	432	204
Middlebury, Addison, Vt	498	270
Middle Granville, Washington, N. Y	442	214
Middletown, Middlesex, Conn	327	99
Middletown, New Castle, Del	134	144
Middletown, Orange, N. Y	294	66
Middletown, Butler, Ohio	581	718
Middletown, Dauphin, Pa	124	186
Midland, Midland, Mich	720	796
Mifflintown, Juniata, Pa	172	232
Milford, New Haven, Conn	295	143
Milford, Kent, Del	182	192
Milford, Worcester, Mass	456	228
Milford, Hillsborough, N. H	479	251
Milford, Caroline, Va	76	304
Millburn, Essex, N. J	227	19
Millbury, Worcester, Mass	426	198
Mill City, Humboldt, Nev	2,745	2,874
Milledgeville, Baldwin, Ga	685	913
Millersburgh, Holmes, Ohio	450	579
Millersburgh, Dauphin, Pa	149	209
Millersville, Lancaster, Pa	125	162
Millerton, Dutchess, N. Y	322	94
Millville, Cumberland, N. J	179	127
Milton, Norfolk, Mass	465	237
Milton, Northumberland, Pa	191	216
Milwaukee, Milwaukee, Wis	859	988
Mineral Point, Iowa, Wis	950	1,079
Minersville, Schuylkill, Pa	194	168
Minneapolis, Hennepin, Minn	1,199	1,328
Minonk, Woodford, Ill	836	974

Place, County, and State.	Distance from Washington.	Distance from New York.
	Miles.	*Miles.*
Mishawaka, Saint Joseph, Ind	700	829
Missouri Valley, Harrison, Iowa	1,240	1,369
Mitchell, Lawrence, Ind	691	871
Moberly, Randolph, Mo	1,043	1,180
Mobile, Mobile, Ala	1,048	1,276
Mohawk, Herkimer, N. Y	455	227
Mokelumne, San Joaquin, Cal	3,076	3,205
Moline, Rock Island, Ill	936	1,065
Monmouth, Warren, Ill	952	1,060
Monongahela City, Washington, Pa	279	474
Monroe, Ouachita, La	1,146	1,374
Monroe, Monroe, Mich	583	712
Monroe, Green, Wis	897	1,023
Monroeville, Huron, Ohio	504	633
Monson, Hampden, Mass	386	158
Montana, Boone, Iowa	1,113	1,242
Monterey, Monterey, Cal	3,255	3,384
Montezuma, Poweshiek, Iowa	1,075	1,204
MONTGOMERY, Montgomery, Ala	862	1,090
Montgomery City, Montgomery, Mo	988	1,132
Monticello, Piatt, Ill	810	947
Monticello, White, Ind	706	844
Monticello, Jones, Iowa	1,005	1,134
Monticello, Sullivan, N. Y	338	111
MONTPELIER, Washington, Vt	532	304
Montrose, Westchester, N. Y	268	40
Montrose, Susquehanna, Pa	294	181
Moorestown, Burlington, N. J	149	99
Morehead City, Carteret, N. C	376	604
Morganton, Burke, N. C	435	663
Morgantown, Monongalia, W. Va	295	511
Morris, Grundy, Ill	831	960
Morrisania, Westchester, N. Y	238	10
Morrison, Whitesides, Ill	896	1,025
Morristown, Morris, N. J	246	32
Morristown, Grainger, Tenn	472	700
Morrisville, Madison, N. Y	423	279
Morrow, Warren, Ohio	551	708
Mound City, Pulaski, Ill	907	1,079
Mount Ayr, Ringgold, Iowa	1,189	1,314
Mount Carmel, Northumberland, Pa	206	176
Mount Carroll, Carroll, Ill	920	1,049
Mount Clemens, Macomb, Mich	648	655
Mount Gilead, Morrow, Ohio	504	640
Mount Holly, Burlington, N. J	158	83
Mount Jackson, Shenandoah, Va	120	348
Mount Joy, Lancaster, Pa	133	170
Mount Morris, Livingston, N. Y	381	357
Mount Pleasant, Henry, Iowa	1,011	1,136
Mount Sterling, Brown, Ill	946	1,083
Mount Sterling, Montgomery, Ky	699	879
Mount Vernon, Mobile, Ala	1,077	1,305
Mount Vernon, Jefferson, Ill	859	1,031
Mount Vernon, Posey, Ind	828	998
Mount Vernon, Linn, Iowa	976	1,105

Place, County, and State.	Distance from Washington.	Distance from New York.
	Miles.	*Miles.*
Mount Vernon, Westchester, N. Y	245	17
Mount Vernon, Knox, Ohio	478	618
Muncie, Delaware, Ind	621	758
Muncy, Lycoming, Pa	206	232
Murfreesborough, Rutherford, Tenn	745	973
Murphysborough, Jackson, Ill	913	1,085
Muscatine, Muscatine, Iowa	992	1,121
Muskegon, Muskegon, Mich	774	878
Mystic, New London, Conn	361	133
Mystic Bridge, New London, Conn	363	135
Nacogdoches, Nacogdoches, Tex	1,519	1,747
Nantucket, Nantucket, Mass	507	279
Napa City, Napa, Cal	3,100	3,229
Naperville, DuPage, Ill	802	931
Napoleon, Henry, Ohio	593	722
Narraguagas, Washington, Me	766	538
Nashua, Chickasaw, Iowa	1,090	1,219
Nashua, Hillsborough, N. H	468	240
Nashville, Washington, Ill	873	1,045
NASHVILLE, Davidson, Tenn	779	1,007
Natchez, Adams, Miss	1,186	1,414
Natchitoches, Natchitoches, La	1,381	1,609
Natick, Middlesex, Mass	448	220
Naugatuck, New Haven, Conn	313	85
Navasota, Grimes, Tex	1,567	1,795
Nebraska City, Otoe, Nebr	1,275	1,400
Neenah, Winnebago, Wis	968	1,089
Negaumee, Marquette, Mich	1,146	1,278
Neosho, Newton, Mo	1,221	1,365
Nevada, Storey, Iowa	1,091	1,220
Nevada, Vernon, Mo	1,185	1,329
Nevada City, Nevada, Cal	3,005	3,134
New Albany, Floyd, Ind	676	856
Newark, New Castle, Del	96	131
Newark, Essex, N. J	218	10
Newark, Wayne, N. Y	426	343
Newark, Licking, Ohio	453	592
New Bedford, Bristol, Mass	443	215
New Berne, Craven, N. C	340	568
Newberry, C. H., Newberry, S. C	554	782
New Braunfels, Comal, Tex	1,748	1,976
New Brighton, Beaver, Pa	332	461
New Britain, Hartford, Conn	332	103
New Brunswick, Middlesex, N. J	195	32
Newburgh, Orange, N. Y	290	62
Newburyport, Essex, Mass	500	272
New Canaan, Fairfield, Conn	273	45
New Castle, New Castle, Del	116	126
Newcastle, Henry, Ind	634	771

Place, County, and State.	Distance from Washington.	Distance from New York.
	Miles.	*Miles.*
New Castle, Lincoln, Me	626	398
Newcastle, Lawrence, Pa	353	482
New Chicago, Neosho, Kans	1,279	1,423
NEW HAVEN, New Haven, Conn	304	76
New Iberia, Iberia, La	1,337	1,565
New Lexington, Perry, Ohio	450	601
New Lisbon, Columbiana, Ohio	377	506
New Lisbon, Juneau, Wis	993	1,122
New London, New London, Conn	354	126
New Milford, Litchfield, Conn	322	94
Newnan, Coweta, Ga	765	993
NEW ORLEANS, Orleans, La	1,188	1,416
New Oxford, Adams, Pa	104	219
New Philadelphia, Tuscarawas, Ohio	405	534
Newport, Campbell, Ky	566	746
Newport, Sullivan, N. H	483	255
Newport, Perry, Pa	150	210
NEWPORT, Newport, R. I	398	170
Newport, Orleans, Vt	597	369
New Richmond, Clement, Ohio	585	765
New Rochelle, Westchester, N. Y	248	20
Newton, Jasper, Iowa	1,094	1,223
Newton, Middlesex, Mass	461	233
Newton, Sussex, N. J	246	73
Newville, Cumberland, Pa	153	213
New York, New York, N. Y	228	0
Niagara Falls, Niagara, N. Y	469	448
Nicholasville, Jessamine, Ky	678	858
Niles, Berrien, Mich	728	845
Niles, Trumbull, Ohio	388	517
Noblesville, Hamilton, Ind	696	833
Nokomis, Montgomery, Ill	854	991
Norfolk, Norfolk, Va	220	448
Normal, McLean, Ill	842	979
Norristown, Montgomery, Pa	156	106
North Adams, Berkshire, Mass	406	178
Northampton, Hampshire, Mass	380	152
North Attleborough, Bristol, Mass	437	209
North Bridgewater, Plymouth, Mass	446	218
North Brookfield, Worcester, Mass	400	172
North Cambridge, Middlesex, Mass	472	244
North Conway, Carroll, N. H	602	374
North East, Erie, Pa	482	492
North Easton, Bristol, Mass	443	216
Northfield, Rice, Minn	1,199	1,328
Northfield, Washington, Vt	524	296
North McGregor, Clayton, Iowa	1,002	1,131
North Platte, Lincoln, Nebr	1,557	1,686
Northumberland, Northumberland, Pa	180	206
North Vernon, Jennings, Ind	637	817
Norwalk, Fairfield, Conn	275	47
Norwalk, Huron, Ohio	499	628
Norwich, New London, Conn	367	139
Norwich, Chenango, N. Y	397	255
Notre Dame, Saint Joseph, Ind	706	835

Place, County, and State.	Distance from Washington.	Distance from New York.
	Miles.	*Miles.*
Nunda, Livingston, N. Y	379	355
Nyack, Rockland, N. Y	256	30
Oakalla, Iroquois, Ill	794	932
Oakland, Alameda, Cal	3,120	3,332
Oakland, Alleghany, Md	207	422
Oberlin, Lorain, Ohio	479	608
Oceola, Clarke, Iowa	1,139	1,264
Oconomowoc, Waukesha, Wis	890	1,019
Oconto, Oconto, Wis	1,043	1,172
Odell, Livingston, Ill	830	968
Odin, Marion, Ill	840	1,012
Ogden City, Weber, Utah	2,298	2,427
Ogdensburgh, Saint Lawrence, N. Y	576	397
Oil City, Venango, Pa	435	482
Okolona, Chickasaw, Miss	912	1,140
Olathe, Johnson, Kans	1,210	1,354
Old Town, Penobscot, Me	722	494
Old Point Comfort, Elizabeth City, Va	223	375
Olean, Cattaraugus, N. Y	418	394
Olivet, Eaton, Mich	680	809
Olney, Richland, Ill	788	960
Olneyville, Providence, R. I	423	195
OLYMPIA, Thurston, Wash	3,253	3,382
Omaha City, Douglas, Nebr	1,266	1,395
Omro, Winnebago, Wis	958	1,087
Onarga, Iroquois, Ill	775	913
Oneida, Knox, Ill	924	1,053
Oneida, Madison, N. Y	462	268
Oneonta, Otsego, N. Y	416	227
Opelika, Lee, Ala	834	1,062
Opelousas, Saint Landry, La	1,393	1,621
Orange, Orange, Tex	1,649	1,877
Orange, Essex, N. J	221	13
Orange, C. H., Orange, Va	88	316
Orangeburgh C. H., Orangeburgh, S. C	558	786
Oregon, Holt, Mo	1,223	1,348
Oregon City, Clackamas, Oreg	3,143	3,272
Orland, Hancock, Me	725	497
Oroville, Butte, Cal	3,086	3,215
Orrville, Wayne, Ohio	427	556
Osage, Mitchell, Iowa	1,118	1,247
Osage Mission, Neosho, Kans	1,240	1,384
Osceola, Clarke, Iowa	1,139	1,264
Osceola Mills, Clearfield, Pa	259	319
Oshkosh, Winnebago, Wis	957	1,086
Oskalooso, Mahaska, Iowa	1,082	1,207
Oswego, Oswego, N. Y	472	326
Ottawa, La Salle, Ill	856	985
Ottawa, Franklin, Kans	1,242	1,386

Place, County, and State.	Distance from Washington.	Distance from New York.
	Miles.	*Miles.*
Ottawa, Putnam, Ohio	584	713
Ottumwa, Wapello, Iowa	1,058	1,183
Ovid, Clinton, Mich	690	766
Ovid, Seneca, N. Y	349	300
Owatonna, Steele, Minn	1,146	1,275
Owego, Tioga, N. Y	334	236
Owensborough, Daviess, Ky	821	1,001
Owosso, Shiawassee, Mich	680	756
Oxford, La Fayette, Miss	939	1,167
Oxford, Chenango, N. Y	388	247
Oxford, Butler, Ohio	603	744
Oxford, Chester, Pa	98	141
Ozaukee, Ozaukee, Wis	889	1,018
Pacific, Franklin, Mo	943	1,087
Paducah, McCracken, Ky	899	1,079
Painesville, Lake, Ohio	473	558
Palmer, Hampden, Mass	382	154
Palmyra, Marion, Mo	1,000	1,125
Palmyra, Wayne, N. Y	417	351
Pana, Christian, Ill	841	978
Paola, Miami, Kans	1,232	1,376
Paris, Edgar, Ill	765	902
Paris, Bourbon, Ky	646	826
Paris, Monroe, Mo	1,019	1,156
Paris, Henry, Tenn	917	1,098
Parkersburgh, Wood, W. Va	358	573
Parker's Landing, Armstrong, Pa	385	514
Parma, Jackson, Mich	648	777
Passaic, Passaic, N. J	239	11
Paterson, Passaic, N. J	243	16
Patterson, Putnam, N. Y	291	63
Paw Paw, Van Buren, Mich	725	854
Pawtucket, Providence, R. I	423	195
Paxton, Ford, Ill	790	928
Peabody, Essex, Mass	482	254
Pecatonica, Winnebago, Ill	879	1,008
Peekskill, Westchester, N. Y	271	43
Pekin, Tazewell, Ill	876	1,005
Pella, Marion, Iowa	1,097	1,222
Pembina, Pembina, Dak	1,608	1,737
Penn Yan, Yates, N. Y	342	318
Pensacola, Escambia, Fla	1,025	1,253
Pent Water, Oceana, Mich	836	918
Peoria, Peoria, Ill	887	995
Perry, Wyoming, N. Y	397	373
Perrysburgh, Wood, Ohio	567	696
Perryville, Cecil, Md	75	150
Perth Amboy, Middlesex, N. J	217	27
Peru, La Salle, Ill	872	1,001

Place, County, and State.	Distance from Washington.	Distance from New York.
	Miles.	*Miles.*
Peru, Miami, Ind	678	807
Petaluma, Sonoma, Cal	3,158	3,287
Peterborough, Hillsborough, N. H	483	255
Petersburgh, Menard, Ill	901	1,038
Petersburgh, Dinwiddie, Va	139	367
Petroleum Centre, Venango, Pa	467	489
Phelps, Ontario, N. Y	381	352
Philadelphia, Philadelphia, Pa	139	87
Phillipsburgh, Centre, Pa	264	324
Phillipsburgh, Warren, N. J	209	75
Phil Sheridan, Wallace, Kans	1,596	1,740
Phœnixville, Chester, Pa	167	116
Pierceton, Kosciusko, Ind	654	783
Pikesville, Baltimore, Md	46	198
Piedmont, Mineral, W. Va	181	396
Pine Bluff, Jefferson, Ark	1,177	1,405
Piqua, Miami, Ohio	560	697
Pit Hole City, Venango, Pa	451	480
Pittsburgh, Allegheny, Pa	303	432
Pittsfield, Pike, Ill	946	1,083
Pittsfield, Berkshire, Mass	386	158
Pittston, Luzerne, Pa	251	156
Placerville, El Dorado, Cal	3,106	3,235
Plainfield, Union, N. J	250	25
Plainwell, Allegan, Mich	717	846
Plano, Kendall, Ill	825	954
Plantsville, Hartford, Conn	325	97
Platteville, Grant, Wis	948	1,077
Plattsburgh, Clinton, N. Y	576	348
Plattsmouth, Cass, Nebr	1,263	1,388
Plattsburgh, Clinton, Mo	1,175	1,300
Pleasant Hill, Cass, Mo	1,154	1,298
Pleasanton, Linn, Kans	1,230	1,374
Pleasantville, Venango, Pa	462	487
Plymouth, Marshall, Ind	687	816
Plymouth, Plymouth, Mass	483	254
Plymouth, Grafton, N. H	556	328
Plymouth, Luzerne, Pa	240	167
Plymouth, Richland, Ohio	498	627
Point Pleasant, Mason, W. Va	440	655
Pollard, Escambia, Ala	976	1,204
Polo, Ogle, Ill	883	1,012
Pomeroy, Meigs, Ohio	432	647
Pond Creek, Bureau, Ill	900	1,029
Pontiac, Livingston, Ill	819	957
Pontiac, Oakland, Mich	649	704
Portage City, Columbia, Wis	945	1,074
Port Byron, Cayuga, N. Y	453	318
Port Chester, Westchester, N. Y	257	29
Port Clinton, Schuylkill, Pa	175	149
Port Deposit, Cecil, Md	79	160
Port Henry, Essex, N. Y	494	266
Port Huron, Saint Clair, Mich	685	618
Port Jervis, Orange, N. Y	314	87
Portland, Middlesex, Conn	329	103

Place, County, and State.	Distance from Washington.	Distance from New York.
	Miles.	*Miles.*
Portland, Cumberland, Me	572	344
Portland, Multnomah, Oreg	3,127	3,256
Port Lavaca, Calhoun, Tex	1,634	1,862
Portsmouth, Rockingham, N. H	520	292
Portsmouth, Scioto, Ohio	491	706
Portsmouth, Norfolk, Va	222	450
Port Townsend, Jefferson, Wash	3,358	3,487
Potosi, Washington, Mo	971	1,115
Potsdam, Saint Lawrence, N. Y	576	397
Pottstown, Montgomery, Pa	173	129
Pottsville, Schuylkill, Pa	190	164
Poughkeepsie, Dutchess, N. Y	304	75
Poultney, Rutland, Vt	448	220
Prairie City, McDonough, Ill	963	1,071
Prairie City, Jasper, Iowa	1,120	1,245
Prairie du Chien, Crawford, Wis	1,001	1,130
Prescott, Yavapai, Ariz	3,013	3,157
Princess Anne, Somerset, Md	225	235
Princeton, Bureau, Ill	878	1,007
Princeton, Gibson, Ind	780	952
Princeton, Mercer, N. J	184	50
Providence, Luzerne, Pa	262	149
PROVIDENCE, Providence, R. I	419	191
Provincetown, Barnstable, Mass	521	293
Provo City, Utah, Utah	2,372	2,501
Pueblo, Pueblo, Colo	1,947	2,091
Pulaski, Oswego, N. Y	474	301
Pulaski, Giles, Tenn	788	1,016
Quincy, Adams, Ill	985	1,122
Quincy, Norfolk, Mass	461	233
Quincy, Branch, Mich	639	768
Racine, Racine, Wis	834	963
Rahway, Union, N. J	207	20
RALEIGH, Wake, N. C	300	528
Randolph, Norfolk, Mass	451	223
Randolph, Cattaraugus, N. Y	455	431
Rantoul, Champaign, Ill	806	943
Ravenna, Portage, Ohio	406	535
Rawling's Springs, Carbon, Wyo	1,975	2,104
Reading, Middlesex, Mass	476	248
Reading, Berks, Pa	155	129
Red Bluff, Tehama, Cal	3,129	3,258
Red Bank, Monmouth, N. J	219	72

Place, County, and State.	Distance from Washington.	Distance from New York.
	Miles.	*Miles.*
Red Oak Junction, Montgomery, Iowa	1,224	1,349
Red River Landing, Point Coupee, La	1,258	1,486
Red Wing, Goodhue, Minn	1,141	1,270
Relay House, (St. Dennis P. O.,) Baltimore, Md	31	199
Reno, Washoe, Nev	2,886	3,015
Renovo, Clinton, Pa	271	296
Rhinebeck, Dutchess, N. Y	318	90
Richfield Springs, Otsego, N. Y	417	275
Richland Centre, Richland, Wis	957	1,086
Richmond, Wayne, Ind	607	744
Richmond, Madison, Ky	811	991
Richmond, Sagadahoc, Me	618	390
Richmond, Ray, Mo	1,144	1,288
Richmond, Richmond, N. Y	242	14
RICHMOND, Henrico, Va	115	343
Ridgeway, Lenawee, Mich	599	728
Ridgway, Elk, Pa	349	374
Rio Grande City, Starr, Tex	2,033	2,261
Ripley, Brown, Ohio	621	801
Ripon, Fond du Lac, Wis	942	1,071
Riverhead, Suffolk, N. Y	307	79
Rochelle, Ogle, Ill	847	976
Rochester, Olmsted, Minn	1,130	1,259
Rochester, Monroe, N. Y	395	371
Rochester, Beaver, Pa	329	458
Rockaway, Morris, N. J	248	41
Rockford, Winnebago, Ill	865	994
Rock Island, Rock Island, Ill	954	1,083
Rockland, Knox, Me	657	429
Rockport, Spencer, Ind	785	965
Rockport, Essex, Mass	500	272
Rockville, Tolland, Conn	362	134
Rockville, Parke, Ind	748	885
Rockville, Montgomery, Md	15	243
Rogersville Junction, Hawkins, Tenn	458	686
Rolla, Phelps, Mo	1,020	1,164
Rome, Floyd, Ga	664	892
Rome, Oneida, N. Y	474	255
Romeo, Macomb, Mich	678	653
Rondout, Ulster, N. Y	320	92
Roseville, Placer, Cal	3,022	3,151
Rouse's Point, Clinton, N. Y	587	359
Rouseville, Venango, Pa	460	486
Roxbury, Litchfield, Conn	325	97
Rushford, Fillmore, Minn	1,084	1,213
Rushville, Rush, Ind	643	780
Rushville, Schuyler, Ill	956	1,069
Russellville, Franklin, Ala	809	1,037
Russellville, Logan, Ky	815	995
Russellville, Jefferson, Tenn	466	694
Rutland, Rutland, Vt	465	237
Rye, Westchester, N. Y	255	27

Place, County, and State.	Distance from Washington.	Distance from New York.
	Miles.	*Miles.*
Sabula, Jackson, Iowa	927	1,056
Sackett's Harbor, Jefferson, N. Y	516	337
Saco, York, Me	559	331
SACRAMENTO CITY, Sacramento, Cal	3,040	3,169
Sag Harbor, Suffolk, N. Y	338	104
Saginaw, Saginaw, Mich	702	778
Saint Albans, Franklin, Vt	566	338
Saint Anthony's Falls, Hennepin, Minn	1,197	1,326
Saint Augustine, Saint John's, Fla	1,000	1,228
Saint Charles, Kane, Ill	810	939
Saint Charles, Winona, Minn	1,108	1,237
Saint Charles, Saint Charles, Mo	927	1,071
Saint Clair, Saint Clair, Mich	680	630
Saint Clair, Schuylkill, Pa	195	169
Saint Clairsville, Belmont, Ohio	366	509
Saint Cloud, Stearne, Minn	1,262	1,391
Saint John's, Clinton, Mich	700	776
Saint Johnsbury, Caledonia, Vt	552	324
Saint Johnsville, Montgomery, N. Y	437	209
Saint Joseph, Berrien, Mich	757	886
Saint Joseph, Buchanan, Mo	1,179	1,316
Saint Mark's, Wakulla, Fla	965	1,193
Saint Mary's, Pottawatomie, Kans	1,281	1,425
Saint Mary's, Auglaize, Ohio	585	714
Saint Mary's, Elk, Pa	339	364
Saint Louis, Saint Louis, Mo	906	1,050
SAINT PAUL, Ramsey, Minn	1,188	1,317
Saint Peter, Nicollet, Minn	1,220	1,350
Salamanca, Cattaraugus, N. Y	437	413
Salem, Marion, Ill	835	1,007
Salem, Washington, Ind	711	891
Salem, Essex, Mass	480	252
Salem, Salem, N. J	183	133
Salem, Washington, N. Y	421	193
Salem, Forsyth, N. C	342	570
Salem, Columbiana, Ohio	370	499
SALEM, Marion, Oreg	3,180	3,309
Salem, Roanoke, Va	239	467
Salina, Saline, Kans	1,376	1,520
Salina, Onodaga, N. Y	438	295
Salisbury, Wicomico, Md	212	222
Salisbury, Rowan, N. C	355	583
SALT LAKE CITY, Salt Lake, Utah	2,335	2,464
San Antonio, Bexar, Tex	1,786	2,012
San Bernardino, San Bernardino, Cal	3,363	3,507
San Diego, San Diego, Cal	3,199	3,343
Sandusky, Erie, Ohio	505	634
Sandwich, DeKalb, Ill	829	958
Sandwich, Barnstable, Mass	460	232
Sandy Hill, Washington, N. Y	429	203
San Francisco, San Francisco, Cal	3,123	3,252
San José, Santa Clara, Cal	3,167	3,296
San Juan Island, Whatcom, Wash	3,961	4,090
Santa Clara, Santa Clara, Cal	3,170	3,299
Santa Cruz, Santa Cruz, Cal	3,205	3,334

Place, County, and State.	Distance from Washington.	Distance from New York.
	Miles.	*Miles.*
Santa Fé, Santa Fé, N. Mex	2,120	2,264
Saratoga Springs, Saratoga, N. Y	411	183
Saugerties, Ulster, N. Y	331	103
Sauk Rapids, Benton, Minn	1,264	1,393
Sault de Ste. Marie, Chippewa, Mich	973	1,028
Savannah, Chatham, Ga	682	910
Savannah, Andrew, Mo	1,194	1,331
Saybrook, Middlesex, Conn	335	107
Schenectady, Schenectady, N. Y	390	162
Schoharie, Schoharie, N. Y	414	186
Schuylkill, Chester, Pa	165	115
Schuylkill Haven, Schuylkill, Pa	186	160
Scranton, Luzerne, Pa	260	147
Seaford, Sussex, Del	193	203
Searsport, Waldo, Me	714	486
Seattle, King, Wash	3,204	3,433
Sedalia, Pettis, Mo	1,095	1,239
Selin's Grove, Snyder, Pa	171	209
Selma, Dallas, Ala	861	1,089
Seneca Falls, Seneca, N. Y	398	334
Seymour, New Haven, Conn	306	78
Seymour, Jackson, Ind	651	831
Shaker Village, Merrimack, N. H	522	294
Shakopee, Scott, Minn	1,204	1,333
Shamburgh, Venango, Pa	464	489
Shamokin, Northumberland, Pa	197	185
Sharon, Mercer, Pa	374	503
Sharon Springs, Schoharie, N. Y	432	204
Sharpsville, Mercer, Pa	377	506
Shasta, Shasta, Cal	3,177	3,306
Shawneetown, Gallatin, Ill	850	1,022
Sheboygan, Sheboygan, Wis	925	1,054
Shelbina, Shelby, Mo	1,020	1,157
Shelburne Falls, Franklin, Mass	416	188
Shelby, Richland, Ohio	500	629
Shelbyville, Shelby, Ill	825	962
Shelbyville, Shelby, Ind	652	800
Shelbyville, Shelby, Ky	679	859
Shelbyville, Bedford, Tenn	731	959
Sheldon, Franklin, Vt	574	348
Shenandoah, Schuylkill, Pa	222	160
Sherburne, Chenango, N. Y	408	266
Sherman, Grayson, Tex	1,537	1,681
Sherman, Albany, Wyo	1,815	1,944
Shippensburgh, Cumberland, Pa	99	224
Shreveport, Caddo, La	1,276	1,504
Sidney, Fremont, Iowa	1,260	1,385
Sidney, Cheyenne, Nebr	1,680	1,809
Sidney, Shelby, Ohio	572	709
Sigourney, Keokuk, Iowa	1,058	1,187
Sing Sing, Westchester, N. Y	260	32
Sioux City, Woodbury, Iowa	1,288	1,418
Sitka, ———, Alaska	4,831	4,960
Skaneateles, Onondaga, N. Y	428	310
Skowhegan, Somerset, Me	672	444

Place, County, and State.	Distance from Washington.	Distance from New York.
	Miles.	*Miles.*
Slatington, Lehigh, Pa	207	109
Smithville, (Renwick P. O.) Lee, Ga	800	1,028
Smyrna, Kent, Del	148	158
Somerset, Somerset, Pa	211	426
Somerville, Somerset, N. J	213	37
Somerville, Fayette, Tenn	910	1,138
Sonora, Tuolumne, Cal	3,122	3,251
South Bend, Saint Joseph, Ind	715	834
South Berwick, York, Me	534	306
Southbridge, Worcester, Mass	421	193
South Deerfield, Franklin, Mass	395	167
South Hadley, Hampshire, Mass	379	151
South Hadley Falls, Hampshire, Mass	376	148
South Haven, Van Buren, Mich	744	873
Southington, Hartford, Conn	326	98
South Norwalk, Fairfield, Conn	273	45
South Pass City, Sweet Water, Wyo	2,224	2,353
South Orange, Essex, N. J	223	15
Sparta, Randolph, Ill	920	1,092
Sparta, Monroe, Wis	1,029	1,158
Spartanburgh C. H., Spartanburgh, S. C	601	829
Spencer, Owen, Ind	727	864
Spencer, Worcester, Mass	403	175
Stafford Springs, Tolland, Conn	393	165
SPRINGFIELD, Sangamon, Ill	871	1,008
Springfield, Hampden, Mass	367	139
Springfield, Greene, Mo	1,147	1,293
Springfield, Clark, Ohio	528	665
Springfield, Windsor, Vt	460	232
Springfield, Walworth, Wis	864	993
Stamford, Fairfield, Conn	265	37
Stapleton, Richmond, N. Y	236	8
Staunton, Augusta, Va	157	385
Steilacoom City, Pierce, Wash	3,264	3,393
Sterling, Whitesides, Ill	882	1,011
Steubenville, Jefferson, Ohio	346	475
Stevenson, Jackson, Ala	665	893
Stillwater, Washington, Minn	1,212	1,341
Stockbridge, Berkshire, Mass	380	152
Stockton, San Joaquin, Cal	3,087	3,216
Stoneham, Middlesex, Mass	474	246
Stonington, New London, Conn	366	138
Stoughton, Dane, Wis	888	1,017
Stowe, Lamoille, Vt	553	324
Strasburgh, Shenandoah, Va	96	324
Stratford, Fairfield, Conn	290	62
Streator, La Salle, Ill	868	997
Stroudsburgh, Monroe, Pa	244	96
Sturgis, Saint Joseph, Mich	670	799
Suffield, Hartford, Conn	357	129
Suffolk, Nansemond, Va	197	425
Suison City, Solano, Cal	3,085	3,214
Sullivan, Sullivan, Ind	772	909
Sumter C. H., Sumter, S. C	512	740
Sunbury, Northumberland, Pa	178	204

Place, County, and State.	Distance from Washington.	Distance from New York.
	Miles.	*Miles.*
Suncook, Merrimack, N. H	498	270
Superior, Douglas, Wis	1,351	1,480
Suspension Bridge, Niagara, N. Y	470	446
Susquehanna Depot, Susquehanna, Pa	320	192
Sycamore, DeKalb, Ill	832	961
Syracuse, Onondaga, N. Y	436	293
Talladega, Talladega, Ala	752	980
TALLAHASSEE, Leon, Fla	954	1,182
Tallmadge, Summit, Ohio	419	548
Tama City, Tama, Iowa	1,043	1,172
Tamaqua, Schuylkill, Pa	195	158
Tampa, Hillsborough, Fla	1,109	1,337
Tarborough, Edgecombe, N. C	266	494
Tarr Farm, Venango, Pa	464	488
Tarrytown, Westchester, N. Y	255	27
Taunton, Bristol, Mass	432	205
Taylorville, Christian, Ill	859	996
Tebeanville, Ware, Ga	779	1,007
Tecumseh, Lenawee, Mich	594	723
Terre Haute, Vigo, Ind	746	883
Terrysville, Litchfield, Conn	331	104
The Dalles, Wasco, Oreg	3,007	3,136
Thibodeaux, La Fourche, La	1,244	1,472
Thomaston, Litchfield, Conn	329	101
Thomaston, Knox, Me	653	425
Thomasville, Thomas, Ga	882	1,110
Thompsonville, Hartford, Conn	358	130
Thorntown, Boone, Ind	711	848
Three Rivers, Saint Joseph, Mich	694	823
Thurlow, Delaware, Pa	121	105
Tidioute, Warren, Pa	421	446
Tiffin, Seneca, Ohio	536	665
Tilton, Belknap, N. H	523	295
Tippecanoe City, Miami, Ohio	572	709
Tipton, Cedar, Iowa	973	1,102
Tipton, Moniteau, Mo	1,068	1,212
Tiskilwa, Bureau, Ill	894	1,023
Titusville, Crawford, Pa	453	481
Toledo, Tama, Iowa	1,046	1,175
Toledo, Lucas, Ohio	558	687
Tomah, Monroe, Wis	1,012	1,141
Tonawanda, Erie, N. Y	452	428
Tonica, La Salle, Ill	880	1,009
Toulon, Stark, Ill	924	1,032
TOPEKA, Shawnee, Kans	1,258	1,402
Towanda, Bradford, Pa	291	229
Townsend, New Castle, Del	138	148
Traverse City, Grand Traverse, Mich	907	1,006
Tremont, Westchester, N. Y	239	13

Place, County, and State.	Distance from Washington.	Distance from New York.
	Miles.	*Miles.*
TRENTON, Mercer, N. J	170	58
Trenton, Gibson, Tenn	928	1,156
Troy, Rensselaer, N. Y	380	152
Troy, Miami, Ohio	568	705
Troy, Bradford, Pa	271	297
Trumansburgh, Tompkins, N. Y	376	278
TUCSON, Pima, Ariz	2,673	2,817
Tullahoma, Coffee, Tenn	708	936
Tunkhannock, Wyoming, Pa	274	175
Tupelo, Lee, Miss	894	1,122
Tuscaloosa, Tuscaloosa, Ala	824	1,052
Tuscola, Douglas, Ill	809	946
Tuscumbia, Colbert, Ala	791	1,019
Tuskegee, Macon, Ala	865	1,093
Tyler, Smith, Tex	1,382	1,610
Tyrone, Blair, Pa	240	300
Uhricksville, Tuscarawas, Ohio	396	525
Umatilla, Umatilla, Oreg	2,897	3,026
Unadilla, Otsego, N. Y	399	244
Union City, Randolph, Ind	591	728
Union City, Obion, Tenn	937	1,131
Union Mills, Erie, Pa	441	466
Union Springs, Cayuga, N. Y	409	335
Uniontown, Fayette, Pa	260	475
Unionville, Union, S. C	572	800
University of Virginia, Albemarle, Va	119	347
Upper Alton, Madison, Ill	920	1,057
Upper Marlborough, Prince George's, Md	20	226
Upper Sandusky, Wyandot, Ohio	521	650
Urbana, Champaign, Ill	791	928
Urbana, Champaign, Ohio	534	671
Utica, Oneida, N. Y	451	240
Uxbridge, Worcester, Mass	431	203
Valatie, Columbia, N. Y	368	140
Vallejo, Solano, Cal	3,100	3,229
Valparaiso, Porter, Ind	727	856
Van Buren, Crawford, Ark	1,286	1,514
Vancouver, Clark, Wash	3,109	3,238
Vandalia, Fayette, Ill	844	981
Van Wert, Van Wert, Ohio	591	720
Vergennes, Addison, Vt	512	284
Versailles, Woodford, Ky	687	867
Vevay, Switzerland, Ind	620	800

Place, County, and State.	Distance from Washington.	Distance from New York.
	Miles.	*Miles.*
Vicksburgh, Warren, Miss	1,061	1,289
Victor, Iowa, Iowa	1,051	1,105
Victor, Ontario, N. Y	376	352
Victoria, ———, Vancouver's I'd	3,427	3,556
Vincennes, Knox, Ind	756	928
Vineland, Cumberland, N. J	175	123
Vinita, ———, Ind. T	1,270	1,414
Vinton, Benton, Iowa	1,017	1,146
Virden, Macoupin, Ill	892	1,029
Virginia City, Madison, Mont	2,643	2,772
Virginia City, Storey, Nev	2,908	3,638
Visalia, Tulare, Cal	3,252	3,381
Volcano, Amador, Cal	3,120	3,249
Wabash, Wabash, Ind	664	793
Wabashaw, Wabashaw, Minn	1,113	1,242
Waco, McLennan, Tex	1,585	1,813
Wadsworth, Medina, Ohio	437	566
Wakefield, Middlesex, Mass	474	246
Waldoborough, Lincoln, Me	638	410
Walla-Walla, Walla Walla, Wash	2,850	2,979
Wallingford, New Haven, Conn	316	88
Waltham, Middlesex, Mass	474	246
Wanatah, Laporte, Ind	718	847
Wapakonetta, Auglaize, Ohio	576	705
Ware, Hampshire, Mass	394	166
Warner, Merrimack, N. H	524	296
Warren, Jo Daviess, Ill	927	1,056
Warren, Trumbull, Ohio	384	513
Warren, Warren, Pa	401	426
Warren, Worcester, Mass	392	164
Warren, Bristol, R. I	431	203
Warrensburgh, Johnson, Mo	1,124	1,268
Warrenton, Warren, N. C	231	459
Warrenton, Fauquier, Va	57	285
Warrenton Junction, (Owl Run P. O.) Fauquier, Va	48	276
Warsaw, Hancock, Ill	973	1,111
Warsaw, Kosciusko, Ind	663	792
Warsaw, Benton, Mo	1,142	1,286
Warsaw, Wyoming, N. Y	398	374
Warwick, Orange, N. Y	290	63
Washington, Wilkes, Ga	667	895
Washington, Tazewell, Ill	845	983
Washington, Daviess, Ind	737	917
Washington, Washington, Iowa	1,030	1,159
Washington, Saint Landry, La	1,400	1,628
Washington, Franklin, Mo	961	1,105
Washington, Warren, N. J	223	71
Washington, Beaufort, N. C	322	550
Washington, Guernsey, Ohio	395	547

Place, County, and State.	Distance from Washington.	Distance from New York.
	Miles.	*Miles.*
Washington, Washington, Pa	334	463
Washington C. H., Fayette, Ohio	518	677
Washington Junction, Baltimore, Md	31	199
Waterbury, New Haven, Conn	320	92
Waterbury, Washington, Vt	562	334
Waterford, Saratoga, N. Y	385	156
Waterford, Erie, Pa	448	473
Waterloo, DeKalb, Ind	650	776
Waterloo, Black Hawk, Iowa	1,055	1,184
Waterloo, Seneca, N. Y	394	337
Watertown, Middlesex, Mass	471	242
Watertown, Jefferson, N. Y	506	327
Watertown, Jefferson, Wis	902	1,031
Water Valley, Yalabusha, Miss	956	1,184
Waterville, Kennebec, Me	654	426
Waterville, Oneida, N. Y	430	288
Wathena, Doniphan, Kans	1,185	1,322
Watkins, Schuyler, N. Y	319	295
Watseka, Iroquois, Ill	757	895
Waukegan, Lake, Ill	808	937
Waukesha, Waukesha, Wis	880	1,009
Waupaca, Waupaca, Wis	993	1,122
Waupun, Fond du Lac, Wis	927	1,056
Wausau, Marathon, Wis	1,059	1,188
Wauseon, Fulton, Ohio	591	720
Waverly, Bremer, Iowa	1,066	1,195
Waverly, Tioga, N. Y	315	255
Waynesborough, Franklin, Pa	110	253
Waynesburgh, Greene, Pa	300	524
Webster, Worcester, Mass	408	180
Webster City, Hamilton, Iowa	1,034	1,263
Weedsport, Cayuga, N. Y	458	315
Weldon, Halifax, N. C	203	431
Wellington, Lorain, Ohio	480	609
Wellsborough, Tioga, Pa	307	328
Wellsville, Allegany, N. Y	381	358
Wellsville, Columbiana, Ohio	351	480
Wenona Station, Marshall, Ill	846	984
Westborough, Worcester, Mass	434	206
West Chester, Chester, Pa	138	119
West Cornwall, Litchfield, Conn	348	120
West Eau Claire, Eau Claire, Wis	1,095	1,224
Westerly, Washington, R. I	372	144
Westfield, Hampden, Mass	365	137
Westfield, Chautauqua, N. Y	500	476
West Killingly, Windham, Conn	393	165
West Liberty, Muscatine, Iowa	1,175	1,304
West Liberty, Logan, Ohio	544	681
West Meriden, New Haven, Conn	323	95
Westminster, Carroll, Md	74	226
West Newton, Middlesex, Mass	450	222
Weston, Platte, Mo	1,222	1,366
West Point, Troup, Ga	812	1,040
West Point, Lowndes, Miss	941	1,169
West Point, Orange, N. Y	280	52

Place, County, State.	Distance from Washington.	Distance from New York.
	Miles.	*Miles.*
Westport, Fairfield, Conn	276	48
West Randolph, Orange, Vt	524	296
West Rutland, Rutland, Vt	462	234
West Troy, Albany, N. Y	379	152
West Union, Fayette, Iowa	1,071	1,200
West Winsted, Licthfield, Conn	349	121
WHEELING, Ohio, W. Va	354	497
Whitehall, Washington, N. Y	450	222
White Haven, Luzerne, Pa	245	147
White Pigeon, St. Joseph, Mich	683	812
White Plains, Westchester, N. Y	253	25
White River Junction, Windsor, Vt	491	263
White Sulphur Springs, Greenbrier, W. Va	248	476
Whitewater, Walworth, Wis	882	1,011
Wickenburgh, Yavapai, Ariz	2,947	3,091
Wilbraham, Hampden, Mass	376	148
Wilkesbarre, Luzerne, Pa	244	158
Williamsburgh, Kings, N. Y	232	4
Williamsport, Lycoming, Pa	219	244
Williamstown, Berkshire, Mass	411	183
Willimantic, Windham, Conn	373	145
Willoughby, Lake, Ohio	463	569
Wilmington, Los Angeles, Cal	3,317	3,461
Wilmington, New Castle, Del	109	119
Wilmington, Will, Ill	825	954
Wilmington, New Hanover, N. C	365	593
Wilmington, Clinton, Ohio	516	699
Winchester, Worcester, Mass	466	238
Winchester, Scott, Ill	925	1,062
Winchester, Randolph, Ind	600	737
Winchester, Clarke, Ky	664	844
Winchester, Middlesex, Mass	424	244
Winchester, Franklin, Tenn	701	929
Winchester, Frederick, Va	115	303
Windsor, Windsor, Vt	477	250
Windsor Locks, Hartford, Conn	353	125
Winnemucca, Humboldt, Nev	2,717	2,846
Winnsborough, Fairfield, S. C	470	698
Winona, Winona, Minn	1,080	1,209
Winooski Falls, Chittenden, Vt	534	306
Winsted, Litchfield, Conn	348	120
Winterport, Waldo, Me	726	498
Winterset, Madison, Iowa	1,171	1,300
Winthrop, Kennebec, Me	629	401
Wiscasset, Lincoln, Me	619	391
Woburn, Middlesex, Mass	474	246
Wolcottville, Litchfield, Conn	339	111
Wolfborough, Carroll, N. H	573	345
Woodstock, McHenry, Ill	823	952
Woodstock, Shenandoah, Va	109	337
Woodstock, Windsor, Vt	495	267
Woodville, Wilkinson, Miss	1,224	1,452
Woonsocket Falls, Providence, R. I	435	207
Wooster, Wayne, Ohio	438	567
Worcester, Worcester, Mass	422	194

Place, County, and State.	Distance from Washington.	Distance from New York.
	Miles.	*Miles.*
Wyalusing, Bradford, Pa	307	225
Wyandotte, Wyandotte, Kan	1,192	1,336
Wyandotte, Wayne, Mich	606	695
Wytheville, Wythe, Va	312	540
Xenia, Greene, Ohio	542	679
Yankee Ridge, Coshocton, Ohio	450	579
YANKTON, Yankton, Dak	1,349	1,479
Yazoo City, Yazoo, Miss	1,078	1,306
Yellow Springs, Greene, Ohio	551	688
Yonkers, Westchester, N. Y	245	16
York, York, Pa	96	189
Yorkville, York, S. C	465	693
Youngstown, Niagara, N. Y	481	457
Youngstown, Mahoning, Ohio	370	499
Ypsilanti, Washtenaw, Mich	621	708
Yreka, Siskiyou, Cal	3,292	3,421
Zanesville, Muskingum, Ohio	428	586

LIST

OF THE

MILITARY POSTS OF THE UNITED STATES,

WITH

THEIR DISTANCES TO ADJACENT POSTS AND OTHER POINTS,

AS COMPUTED IN THE

OFFICE OF THE QUARTERMASTER-GENERAL, U. S. A.

LIST OF THE MILITARY POSTS OF THE UNITED STATES, WITH THEIR DISTANCES TO ADJACENT POSTS AND OTHER POINTS.

ALBUQUERQUE, N. MEX., TO—

	Miles.
Camp Apache, Maricopa, Ariz	259
Cheyenne, Laramie, Wyo	666
Denver, Arapahoe, Colo	564
El Paso, El Paso, Tex	289
Fort Bascom, San Miguel, N. Mex	322
Fort Bayard, Grant, N. Mex	344
Fort Bliss, El Paso, Tex	292
Fort Craig, Socorro, N. Mex	118
Fort Cummings, Grant, N. Mex	299
Fort Garland, Costilla, Colo	234
Fort Dodge, Ford, Kans	792
Fort Leavenworth, Leavenworth, Kans	941
Fort Lyon, Bent, Colo	311
Fort McRae, Socorro, N. Mex	150
Fort Reynolds, Pueblo, Colo	403
Fort Seldon, Doña Aña, N. Mex	218
Fort Stanton, Socorro, N. Mex	168
Fort Sumner, San Miguel, N. Mex	196
Fort Union, Mora, N. Mex	177
Fort Wingate, N. Mex., (old,)	85
Fort Wingate, N. Mex., (new,)	137
Grenada, Colo	440
Kansas City, Jackson, Mo	1,002
Kit Carson, Greenwood, Colo	513
Mesilla, Doña Aña, N. Mex	239
Santa Fé, Santa Fé, N. Mex	71
Topeka, Kans	873
Washington, Washington, D. C	2,129
Zuni, N. Mex	139

APALACHICOLA, FLA., TO—

Cedar Keys, Levy, Fla	168
Fort Barrancas, Escambia, Fla	158
Fort Jefferson, Monroe, Fla	562
Key West, Monroe, Fla	498
New Orleans, Orleans, La	390
Pensacola, Escambia, Fla	150
St. Mark's, Wakulla, Fla	72
Washington, Washington, D. C	1,054

ASTORIA, OREG., TO—

	Miles.
Boisé City, Ada, Idaho	596
Camp Barney, Grant, Oreg	471
Camp McDermitt, Humboldt, Nev	796
Camp Three Forks, Owyhee, Idaho	701
Camp Warner, Grant, Oreg	601
Canyon City, Grant, Oreg	396
Cape Disappointment, Pacific, Wash	15
Eugene City, Lane, Oreg	210
Fort Boisé, Ada, Idaho	596
Fort Colville, Stevens, Wash	624
Fort Klamath, Wasco, Oreg	489
Fort Lapwai, Nez Percés, Idaho	468
Fort Steilacoom, Pierce, Wash	167
Fort Stevens, Clatsop, Oreg	8
Jacksonville, Jackson, Oreg	389
Lewiston, Nez Percés, Idaho	456
Monticello, Cowlitz, Wash	56
Olympia, Thurston, Wash	141
Portland, Multnomah, Oreg	96
Port Townsend, Jefferson, Wash	273
Red Bluff, Tehama, Cal	576
Reno, Washoe, Nev	832
Sacramento, Sacramento, Cal	714
Sacramento Junction, Placer, Cal	696
San Francisco, San Francisco, Cal	642
San Juan Island, Whatcom, Wash	303
Sitka, Alaska	975
The Dalles, Wasco, Oreg	216
Umatilla, Umatilla, Oreg	316
Vancouver, Clarke, Wash	114
Victoria, Brit. Col	353
Walla Walla, Walla Walla, Wash	371
Wallula, Walla Walla, Wash	341
Washington, Washington, D. C	3,728

ATCHISON, KANS., TO—

Denver, Arapahoe, Colo	620
Dodge City, Kans	348
Emporia Junction, Lyon, Kans	110
Emporia, Lyon, Kans	111
Fort Dodge, Kans	352
Fort Gibson, C. N., Ind. Ter	313
Fort Harker, Ellsworth, Kans	200
Fort Hays, Kans	170
Fort Larned, Pawnee, Kans	296
Fort Leavenworth, Leavenworth, Kans	18
Fort Lyon, Bent, Colo	519
Grenada, Colo	279
Kansas City, Jackson, Mo	48
Kit Carson, Greenwood, Colo	469
Larned, Kans	292
North Topeka, Shawnee, Kans	49
Saint Louis, St. Louis, Mo	330
Topeka, Shawnee, Kans	50
Washington, Washington, D. C	1,237
Waterville, Marshall, Kans	100

AUSTIN, TEXAS, TO—

	Miles.
Belton, Bell, Tex	58
Brashear City, St. Mary's, La	435
Brazos Santiago, Cameron, Tex	336
Bremond, Robertson, Tex	114
Brenham, Washington, Tex	89
Brownsville, Cameron, Tex	369
Cameron, Milam, Tex	74
Columbia, Brazoria, Tex	214
Columbus, Colorado, Tex	98
Corpus Christi, Nueces, Tex	315
Corsicana, Navarro, Tex	157
Denison, Texas	401
El Paso, El Paso, Tex	781
Fort Bliss, El Paso, Tex	778
Fort Chadbourne, Runnels, Tex	351
Fort Clark, Kinney, Tex	203
Fort Concho, Bexar, Tex	306
Fort Davis, Presidio, Tex	556
Fort Duncan, Maverick, Tex	248
Fort Gibson, Cherokee N., Ind. Ter	573
Fort Leavenworth, Leavenworth, Kans	849
Fort McIntosh, Webb, Tex	242
Fort McKavett, Menard, Tex	252
Fort Quitman, El Paso, Tex	696
Fort Sill, Choctaw N., Ind. Ter	597
Fort Stockton, Pecos, Tex	476
Galveston, Galveston, Tex	215
Georgetown, Williamson, Tex	25
Hearne, Texas	185
Hempstead, Austin, Tex	114
Hillsborough, Hill, Tex	133
Houston, Harris, Tex	165
Indianola, Calhoun, Tex	182
Kansas City, Jackson, Mo	820
Laredo, Webb, Tex	242
Longview, Upshur, Tex	360
Marshall, Harrison, Tex	384
Monroe, Ouachita, La	468
Nacogdoches, Nacogdoches, Tex	287
Navasota, Grimes, Tex	134
New Braunfels, Comal, Tex	46
New Orleans, Orleans, La	517
Orange, Orange, Tex	274
Palestine, Texas	276
Ringgold Barracks, Starr, Tex	383
Rio Grande City, Starr, Tex	382
San Antonio, Bexar, Tex	77
Shreveport, Caddo, La	426
Sulphur Springs, (Bright Star,) Hopkins, Tex	260
Vicksburgh, Warren, Miss	620
Victoria, Victoria, Tex	136
Waco, McLennan, Tex	97
Washington, Washington, D. C	1,706

BEAVER CITY, UTAH, TO—

	Miles.
Draper City, Utah	200
Ogden, Utah	261
Omaha, Nebr	1,290
Salt Lake City, Utah	222
Washington, D. C	2,559

BELTON, TEX., TO—

Calvert, Robertson, Tex	71
Corsicana, Navarro, Tex	101
Dallas, Dallas, Tex	141
Fort Richardson, Jack, Tex	198
Fredericksburgh, Gillespie, Tex	143
Galveston, Galveston, Tex	252
Georgetown, Williamson, Tex	33
Houston, Harris, Tex	201
Little Rock, Pulaski, Ark	538
New Braunfels, Comal, Tex	104
San Antonio, Bexar, Tex	135
Shreveport, Caddo, La	292
Waco, McLennan, Tex	39
Washington, Washington, D. C	1,547
Weatherford, Parker, Tex	157

BENICIA ARSENAL, CAL., TO—

Camp Apache, Maricopa, Ariz	1,261
Camp Bidwell, Cal	436
Camp Bowie, Pima, Ariz	1,179
Camp Date Creek, Yavapai, Ariz	894
Camp Gaston, Klamath, Cal	350
Camp Grant, Pima, Ariz	1,033
Camp Halleck, Elko, Nev	609
Camp Hualpai, (late Toll-Gate,) Yavapai, Ariz	913
Camp Independence, Inyo, Cal	552
Camp Lowell, (new,) Pima, Ariz	1,073
Camp McDermitt, Humboldt, Nev	492
Camp McDowell, Maricopa, Ariz	976
Camp Mohave, Mohave, Ariz	777
Camp Verde, Yavapai, Ariz	1,001
Drum Barracks, (Wilmington,) Los Angeles, Cal	456
Ehrenberg, Yuma, Ariz	747
Elko, Elko, Nev	556
Fort Bidwell, Siskiyou, Cal	504
Fort Churchill, Lyon, Nev	274
Fort Whipple, (Prescott,) Yavapai, Ariz	958
Fort Wright, Mendocino, Cal	197
Fort Yuma, San Diego, Cal	742
Los Angeles, Los Angeles, Cal	475
Ogden City, Weber, Utah	831
Omaha City, Douglas, Nebr	1,844
Sacramento, Sacramento, Cal	88
San Bernardino, San Bernardino, Cal	537

BENICIA ARSENAL, CAL., TO—

	Miles.
San Diego, San Diego, Cal	545
San Francisco, San Francisco, Cal	28
Tucson, Pima, Ariz	1,067
Vallejo, Solano, Cal	7
Washington, Washington, D. C	3,105
Wickenburgh, Yavapai, Ariz	867
Wilmington, (Drum Barracks,) Los Angeles, Cal	455

BISMARCK, DAK., TO—

Fort Abercrombie, Shyenne, Dak	232
Fort Abraham Lincoln, Dak	4
Fort Benton, Choteau, Mont	975
Fort Buford, Dak	245
Fort Pembina, Pembina, Dak	350
Fort Randall, Dak	471
Fort Rice, Dak	28
Fort Seward, Dak	100
Fort Snelling, Hennepin, Minn	463
Fort Stevenson, Dak	95
Fort Sully, Buffalo, Dak	286
Fort Thompson, (Crow Creek,) Dak	361
Mendota, Dakota, Minn	465
Minneapolis, Hennepin, Minn	450
Morehead, Minn	198
Pembina, Pembina, Dak	347
St. Paul, Ramsey, Minn	475
Sioux City, Woodbury, Iowa	602
Washington, Washington, D. C	1,663
Yankton, Yankton, Dak	542

BOISÉ CITY, IDAHO, TO—

Camp Bidwell, Siskiyou, Cal	703
Camp Gaston, Klamath, Cal	875
Camp Halleck, Elko, Nev	454
Camp Harney, Grant, Oreg	275
Camp Independence, Inyo, Cal	737
Camp Lincoln, ———, Cal	985
Camp McDermitt, Humboldt, Nev	200
Camp Three Forks, Owyhee, Idaho	105
Camp Warner, Grant, Oreg	405
Canyon City, Grant, Oreg	200
Cape Disappointment, Pacific, Wash	611
Corinne, Box Elder, Utah	317
Dalles (The), Wasco, Oreg	380
Elko, Elko, Nev	424
Fort Boisé, Ada, Idaho	½
Fort Colville, Stevens, Wash	524
Fort Hall, Oneida, Idaho	457
Fort Klamath, Wasco, Oreg	535
Fort Leavenworth, Leavenworth, Kans	1,554
Fort Lapwai, Nez Percés, Idaho	368

BOISÉ CITY, IDAHO, TO—

	Miles.
Fort Stevens, Clatsop, Oreg	604
Kelton, Box Elder, Utah	250
Ogden City, Weber, Utah	342
Olympia, Thurston, Wash	638
Omaha City, Douglas, Nebr	1,371
Portland, Multnomah, Oreg	500
Reno, Washoe, Nev	450
Sacramento, Sacramento, Cal	604
San Juan Island, Whatcom, Wash	752
San Francisco, San Francisco, Cal	687
Silver City, Owyhee, Idaho	70
Sitka, Alaska	1,575
Umatilla, Umatilla, Oreg	326
Vancouver, Clarke, Wash	432
Victoria, Brit. Col	732
Walla Walla, Walla Walla, Wash	271
Wallula, Walla Walla, Wash	301
Washington, Washington, D. C	2,640
Winnemucca, Humboldt, Nev	280

BOSTON, TEX., TO—

Clarksville, Red River, Tex	40
Corsicana, Navarro, Tex	199
Fort Richardson, Jack, Tex	200
Galveston, Galveston, Tex	461
Houston, Harris, Tex	410
Jefferson, Marion, Tex	49
Little Rock, Pulaski, Ark	207
Marshall, Harrison, Tex	65
Memphis, Shelby, Tenn	342
New Orleans, Orleans, La	529
San Antonio, Bexar, Tex	433
Sherman, Grayson, Tex	140
Shreveport, Caddo, Tex	107
Vicksburgh, Warren, Miss	301
Washington, Hempstead, Ark	67
Washington, D. C	1,279

BOZEMAN, MONT., TO—

Corinne, Box Elder, Utah	453
Deer Lodge City, Deer Lodge, Mont	158
Fort Benton, Choteau, Mont	247
Fort Colville, Stevens, Wash	606
Fort Ellis, Gallatin, Mont	3
Fort Shaw, Lewis and Clarke, Mont	183
Gallatin, Gallatin, Mont	35
Gaffney, Madison, Mont	103
Helena, Lewis and Clarke, Mont	103
Missoula, Missoula, Mont	243
Spokan Bridge, Stevens, Wash	513
Virginia City, Madison, Mont	75
Washington, D. C	2,775

BRAZOS SANTIAGO, TEX., TO—

	Miles.
Bremond, Robertson, Tex	439
Brenham, Washington, Tex	371
Brownsville, Cameron, Tex	33
Corpus Christi, Nueces, Tex	193
Denison, Texas	600
Eagle Pass, (Ft. Duncan,) Maverick, Tex	415
Fort Arbuckle, Chickasaw N., Ind. Ter	761
Fort Bliss, El Paso, Tex	1,056
Fort Clark, Kinney, Tex	465
Fort Concho, Bexar, Tex	584
Fort Davis, Presidio, Tex	834
Fort Duncan, (Eagle Pass,) Maverick, Tex	415
Fort Gibson, Cherokee N., Ind. Ter	781
Fort Griffin, Shackelford, Tex	729
Fort McIntosh, Webb, Tex	290
Fort McKavett, Menard, Tex	530
Fort Quitman, El Paso, Tex	974
Fort Sill, Choctaw N., Ind. Ter	796
Fort Stockton, Pecos, Tex	754
Galveston, Galveston, Tex	245
Hempstead, Austin, Tex	346
Houston, Harris, Tex	296
Indianola, Calhoun, Tex	326
Laredo, (Ft. McIntosh,) Webb, Tex	290
New Orleans, Orleans, La	547
Orange, Orange, Tex	406
Point Isabel, Cameron, Texas	5
Ringgold Barracks, (Rio Grande City,) Starr, Tex	150
Rio Grande City, (Ringgold Barracks,) Starr, Tex	150
San Antonio, Bexar, Tex	355
Sherman, Grayson, Tex	632
Victoria, Victoria, Tex	372
Washington, D. C	1,743

BRECKINRIDGE, MINN., TO—

Fort Abercrombie, Shyenne, Dak	14
Fort Abraham Lincoln, Dak	247
Fort Benton, Choteau, Mont	1,210
Fort Buford, Howard, Dak	480
Fort Garry, Brit. Amer	281
Fort Pembina, Pembina, Dak	200
Fort Randall, Todd, Dak	634
Fort Ransom, Ransom, Dak	82
Fort Seward, Dak	144
Fort Snelling, Hennepin, Minn	219
Fort Stevenson, Stevens, Dak	339
Fort Totten, Ramsey, Dak	170
Fort Wadsworth, Stone, Dak	89
Georgetown, Clay, Minn	66
Holy Cross, Clay, Minn	38
Lake Winnepeg, Brit. Amer	314
McCauleyville, Wilkin, Minn	13
Mendota, Dakota, Minn	221
Minneapolis, Hennepin, Minn	214
Moorhead, Clay, Minn	46

BRECKINRIDGE, MINN., TO—

	Miles.
Pembina, Pembina, Dak	205
Pomme de Terre, Grant, Minn	47
Saint Joseph, Pembina, Dak	243
Saint Paul, Ramsey, Minn	213
Sioux City, Woodbury, Iowa	503
Stone Fort, Brit. Amer	291
Washington, D. C	1,446
Yankton, Yankton, Dak	564

BREMOND, TEXAS, TO—

Brenham, Washington, Tex	118
Bright Star, Hopkins, Tex	171
Brownsville, Cameron, Tex	472
Columbia, Brazoria, Tex	193
Columbus, Colorado, Tex	227
Corpus Christi, Nueces, Tex	457
Corsicana, Navarro, Tex	68
Dallas, Dallas, Tex	129
Doublehorn, Burnet, Tex	147
Eagle Pass, Maverick, Tex	362
El Paso, El Paso, Tex	833
Fort Bliss, El Paso, Tex	830
Fort Clark, Kinney, Tex	240
Fort Concho, Bexar, Tex	358
Fort Davis, Presidio, Tex	608
Fort Duncan, Maverick, Tex	362
Fort McIntosh, Webb, Tex	356
Fort McKavett, Menard, Tex	303
Fort Quitman, El Paso, Tex	748
Fort Richardson, Jack, Tex	205
Fort Stockton, Pecos, Tex	528
Fredericksburgh, Gillespie, Tex	199
Galveston, Galveston, Tex	194
Georgetown, Williamson, Tex	89
Hempstead, Austin, Tex	93
Hillsborough, Hill, Tex	83
Houston, Harris, Tex	143
Indianola, Calhoun, Tex	296
Laredo, Webb, Tex	356
Marlin, Falls, Tex	18
Marshall, Harrison, Tex	218
Monroe, Ouachita, La	380
Navasota, Grimes, Tex	73
New Orleans, Orleans, La	496
Orange, Orange, Tex	253
Ringgold Barracks, Starr, Tex	598
Rio Grande City, Starr, Tex	597
San Antonio, Bexar, Tex	189
Vicksburgh, Warren, Miss	454
Victoria, Victoria, Tex	250
Waco, McLennan, Tex	48
Washington, D. C	1,539
Waxahatchie, Ellis, Tex	98

BRENHAM, TEXAS, TO—

	Miles.
Brownsville, Cameron, Tex	404
Columbia, Brazoria, Tex	125
Columbus, Colorado, Tex	43
Corpus Christi, Nueces, Tex	389
Corsicana, Navarro, Tex	186
Eagle Pass, Maverick, Tex	260
El Paso, El Paso, Tex	870
Fort Bliss, El Paso, Tex	867
Fort Clark, Kinney, Tex	292
Fort Concho, Bexar, Tex	318
Fort Davis, Presidio, Tex	645
Fort McIntosh, Webb, Tex	331
Fort McKavett, Menard, Tex	341
Fort Quitman, El Paso, Tex	785
Fort Sill, Choctaw N., Ind. Ter	515
Fort Stockton, Pecos, Tex	565
Galveston, Galveston, Tex	126
Hallettsville, Lavaca, Tex	77
Hempstead, Austin, Tex	25
Houston, Harris, Tex	75
Indianola, Calhoun, Tex	168
Lampasas, Lampasas, Tex	154
New Orleans, Orleans, La	428
Ringgold Barracks, Starr, Tex	530
Rio Grande City, Starr, Tex	529
San Antonio, Bexar, Tex	166
Victoria, Victoria, Tex	122
Waco, McLennan, Tex	166
Washington, D. C.	1,608

BROWNSVILLE, TEX., TO—

Corpus Christi, Nueces, Tex	160
Denison, Texas	665
Eagle Pass, Maverick, Tex	382
Fort Bliss, El Paso, Tex	1,023
Fort Brown, Cameron, Tex	½
Fort Clark, Kinney, Tex	427
Fort Concho, Bexar, Tex	551
Fort Duncan, Maverick, Tex	382
Fort Gibson, Cherokee N., Ind. Ter	814
Fort Griffin, Shackelford, Tex	696
Fort McIntosh, Webb, Tex	257
Fort McKavett, Menard, Tex	497
Fort Quitman, El Paso, Tex	941
Fort Sill, Choctaw N., Ind. Ter	859
Fort Stockton, Pecos, Tex	721
Galveston, Galveston, Tex	278
Houston, Harris, Tex	329
Indianola, Calhoun, Tex	293
New Orleans, Orleans, La	580
Orange, Orange, Tex	439
Ringgold Barracks, Starr, Tex	117
Rio Grande City, Starr, Tex	117
San Antonia, Bexar, Tex	322
Victoria, Victoria, Tex	339
Washington, D. C	1,770

CAMP APACHE, ARIZ., TO—

	Miles.
Albuquerque, N. M	259
Camp Bowie, Pima, Ariz	362
Camp Goodwin, Pima, Ariz	75
Camp Halleck, Elko, Nev	1,723
Camp Hualpai, Yavapai, Ariz	540
Camp Independence, Inyo, Cal	1,109
Camp Lincoln, ———, Cal	1,626
Camp Lowell, Pima, Ariz	250
Camp McDermitt, Humboldt, Nev	1,629
Camp McDowell, Maricopa, Ariz	333
Camp Mohave, Mohave, Ariz	667
Camp Reno, Maricopa, Ariz	366
Camp Verde, Yavapai, Ariz	544
Camp Wallen, Pima, Ariz	301
Camp Wright, Mendocino, Cal	1,430
Drum Barracks, Los Angeles, Cal	827
Ehrenberg, Yuma, Ariz	534
Elko, Elko, Nev	1,693
Fort Tulerosa, N. M	164
Fort Whipple, (Prescott,) Yavapai, Ariz	505
Fort Wingate, East Santa Aña, N. Mex	150
Fort Yuma, San Diego, Cal	519
Grenada, Col	699
Hardyville, Mohave, Ariz	674
Kit Carson, Greenwood, Colo	802
Los Angeles, Los Angeles, Cal	806
Maricopa Wells, Pima, Ariz	319
Pueblo, Col	642
Sacramento, Sacramento, Cal	1,225
San Bernardino, San Bernardino, Cal	744
San Diego, San Diego, Cal	716
San Francisco, San Francisco, Cal	1,233
Tucson, Pima, Ariz	250
Washington, D. C	2,278
Wickenburgh, Yavapai, Ariz	414
Wilmington, Los Angeles, Cal	826

CAMP BAKER, MONT., TO—

Corinne, Box Elder, Utah	505
Diamond City, Meagher, Mont	20
Fort Benton, Choteau, Mont	204
Fort Colville, Stevens, Wash	568
Fort Ellis, Gallatin, Mont	175
Fort Leavenworth, Leavenworth, Kansas	1,725
Fort Shaw, Lewis and Clarke, Mont	140
Helena, Lewis and Clarke, Mont	60
Missoula, Missoula, Mont	200
Spokan Bridge, Stevens, Wash	470
Virginia City, Madison, Mont	183
Walla Walla, Walla Walla, Wash	625
Washington, D. C	2,828

CAMP BEALE'S SPRINGS, ARIZ., TO—

	Miles.
Camp Hualpai, Yavapai, Ariz	95
Camp Mohave, Mohave, Ariz	40
Fort Yuma, San Diego, Cal	280
Los Angeles, Los Angeles, Cal	352
Sacramento, Sacramento, Cal	771
San Diego, San Diego, Cal	477
San Francisco, San Francisco, Cal	799
Washington, Washington, D. C	3,370
Wilmington, Los Angeles, Cal	372

CAMP BIDWELL, CAL., TO—

Camp Gaston, Klamath, Cal	551
Camp Halleck, Elko, Nev	597
Camp Independence, Inyo, Cal	540
Camp McDermitt, Humboldt, Nev	503
Camp Three Forks, Owyhee, Idaho	651
Camp Warner, Grant, Oreg	50
Camp Wright, Mendocino, Cal	685
Canyon City, Grant, Oreg	255
Cape Disappointment, Pacific, Wash	975
Dallas (The), Wasco, Oreg	984
Elko, Elko, Nev	567
Fort Boisé, (Boisé City,) Ada, Idaho	703
Fort Colville, Stevens, Wash	1,227
Fort Hall, Oneida, Idaho	958
Fort Klamath, Wasco, Oreg	140
Fort Lapwai, Nez Percés, Idaho	1,071
Fort Stevens, Clatsop, Oreg	968
Fort Yuma, San Diego, Cal	1,109
Kelton, Box Elder, Utah	751
Los Angeles, Los Angeles, Cal	769
Marysville, Yuba, Cal	296
Ogden City, Weber, Utah	842
Omaha City, Douglas, Nebr	1,871
Oro, Placer, Cal	266
Portland, Multnomah, Oreg	864
Reno, Washoe, Nev	253
Sacramento, Sacramento, Cal	350
San Francisco, San Francisco, Cal	395
San Juan Island, Whatcom, Wash	1,126
Silver City, Owyhee, Idaho	616
Sitka, Alaska	2,114
Susanville, Lassen, Cal	155
Umatilla, Umatilla, Oreg	1,029
Vancouver, Clarke, Wash	882
Victoria, Brit. Amer	1,146
Walla Walla, Walla Walla, Wash	974
Wallula, Walla Walla, Wash	1,004
Washington, D. C	3,439
Wilmington, Los Angeles, Cal	789
Winnemucca, Humboldt, Nev	423

CAMP BOWIE, ARIZ., TO—

	Miles.
Camp Crittenden, Pima, Ariz	105
Camp Date Creek, Yavapai, Ariz	359
Camp Gaston, Klamath, Cal	1,485
Camp Goodwin, Pima, Ariz	88
Camp Grant, Pima, Ariz	172
Camp Halleck, Elko, Nev	1,641
Camp Hualpai, Yavapai, Ariz	458
Camp Independence, Inyo, Cal	1,027
Camp Lincoln, ———, Cal	1,544
Camp Lowell, Pima, Ariz	106
Camp McDermitt, Humboldt, Nev	1,547
Camp McDowell, Maricopa, Ariz	237
Camp Mohave, Mohave, Ariz	585
Camp Reno, Maricopa, Ariz	270
Camp Verde, Yavapai, Ariz	462
Camp Wright, Mendocino, Cal	1,326
Drum Barracks, Los Angeles, Cal	745
Elko, Elko, Nev	1,702
Ehrenberg, Yuma, Ariz	452
Fort Bayard, Grant, N. Mex	120
Fort Bliss, El Paso, Tex	278
Fort Cummings, Grant, N. Mex	165
Fort Whipple, (Prescott,) Yavapai, Ariz	423
Fort Yuma, San Diego, Cal	427
Los Angeles, Los Angeles, Cal	724
Maricopa Wells, Pima, Ariz	227
Mesilla, Doña Aña, N. Mex	225
Prescott, Yavapai, Ariz	423
Ralston, Grant, N. Mex	50
Sacramento, Sacramento, Cal	1,143
San Bernardino, San Bernardino, Cal	662
San Diego, San Diego, Cal	634
San Francisco, San Francisco, Cal	1151
Tucson, Pima, Ariz	112
Washington, D. C	2,616
Wickenburgh, Yavapai, Ariz	332
Wilmington, Los Angeles, Cal	744

CAMP BROWN, WYO., TO—

Bozeman, Gallatin, Mont	790
Bryan Station, Uintah, Wyo	142
Camp Douglas, Salt Lake, Utah	352
Camp Stambaugh, Sweetwater, Wyo	32
Cheyenne, Laramie, Wyo	484
Colorado City, El Paso, Colo	662
Corinne, Box Elder, Utah	337
Denver, Arapahoe, Colo	586
Fort Bridger, Uintah, Wyo	197
Fort D. A. Russell, Laramie, Wyo	339
Fort Ellis, Gallatin, Mont	793
Fort Fetterman, Albany, Wyo	654
Fort Fred Steele, Carbon, Wyo	306
Fort Garland, Costilla, Colo	803
Fort Hall, Oneida, Idaho	477
Fort Kearney, Kearney, Nebr	814

CAMP BROWN, WYO., TO—

	Miles.
Fort Laramie, Laramie, Wyo	564
Fort Leavenworth, Leavenworth, Kans	1,183
Fort McPherson, Lincoln, Nebr	729
Fort Sanders, (Laramie City,) Albany, Wyo	430
Fort Sedgewick, Weld, Colo	626
Helena, Lewis and Clarke, Mont	782
North Platte, Lincoln, Nebr	729
Ogden City, Weber, Utah	313
Omaha City, Douglas, Nebr	980
Salt Lake City, Salt Lake, Utah	350
San Francisco, San Francisco, Cal	1,139
Shoshone and Bannack Indian Reservation	1½
Washington, D. C	2,611
Virginia City, Madison, Mont	715

CAMP CRITTENDEN, ARIZ., TO—

Camp Date Creek, Yavapai, Ariz	298
Camp Goodwin, Pima, Ariz	146
Camp Grant, Pima, Ariz	111
Camp Halleck, Elko, Nev	1,580
Camp Hualpai, Yavapai, Ariz	397
Camp Independence, Inyo, Cal	968
Camp Lincoln, ——, Cal	1,483
Camp McDowell, Maricopa, Ariz	176
Camp McDermitt, Humboldt, Nev	1,486
Camp Mohave, Mohave, Ariz	524
Camp Reno, Maricopa, Ariz	209
Camp Verde, Yavapai, Ariz	401
Camp Wallen, Pima, Ariz	20
Camp Wright, Mendocino, Cal	1,184
Ehrenberg, Yuma, Ariz	391
Fort Bliss, El Paso, Tex	344
Fort Cummings, Grant, N. Mex	233
Fort Whipple, Yavapai, Ariz	362
Fort Yuma, San Diego, Cal	376
Los Angeles, Los Angeles, Cal	663
Maricopa Wells, Pima, Ariz	176
Mesilla, Doña Ana, N. Mex	293
Prescott, Yavapai, Ariz	262
Sacramento, Sacramento, Cal	1,082
San Bernardino, San Bernardino, Cal	601
San Diego, San Diego, Cal	573
San Francisco, San Francisco, Cal	1,090
Tucson, Pima, Ariz	51
Washington, D. C	2,721
Wickenburgh, Yavapai, Ariz	271
Wilmington, Los Angeles, Cal	683

CAMP DATE CREEK, ARIZ., TO—

Camp Gaston, Klamath, Cal	1,200
Camp Goodwin, Pima, Ariz	366
Camp Grant, Pima, Ariz	213
Camp Halleck, Elko, Nev	1,217

CAMP DATE CREEK, ARIZ., TO—

	Miles.
Camp Hualpai, Yavapai, Ariz	75
Camp Independence, Inyo, Cal	712
Camp Lincoln, ———, Cal	1,259
Camp Lowell, (new,) Pima, Ariz	253
Camp McDermitt, Humboldt, Nev	1,242
Camp McDowell, Maricopa, Ariz	136
Camp Mohave, Mohave, Ariz	280
Camp Reno, Maricopa, Ariz	169
Camp Verde, Yavapai, Ariz	99
Camp Wallen, Pima, Ariz	308
Ehrenberg, Yuma, Ariz	147
Elko, Elko, Nev	1,187
Fort Bliss, El Paso, Tex	598
Fort Cummings, Grant, N. Mex	487
Fort Whipple, (Prescott,) Yavapai, Ariz	60
Fort Yuma, San Diego, Cal	254
Havilah, Kern, Cal	559
Los Angeles, Los Angeles, Cal	419
Maricopa Wells, Pima, Ariz	122
Mesilla, Doña Ana, N. Mex	547
Ogden City, Weber, Utah	912
Phœnix, Maricopa, Ariz	97
Prescott, Yavapai, Ariz	60
Sacramento, Sacramento, Cal	838
San Bernardino, San Bernardino, Cal	357
San Diego, San Diego, Cal	451
San Francisco, San Francisco, Cal	866
Tipton, Cal	594
Tucson, Pima, Ariz	247
Washington, D. C	3,210
Wickenburgh, Yavapai, Ariz	27
Wilmington, Los Angeles, Cal	439

CAMP DOUGLAS, UTAH, TO—

Bozeman City, Gallatin, Mont	516
Bryan Station, Uintah, Wyo	210
Camp Stambaugh, Sweetwater, Wyo	320
Cheyenne, Laramie, Wyo	542
Corinne, Box Elder, Utah	63
Denver, Arapahoe, Colo	644
Fort Bridger, Uintah, Wyo	177
Fort D. A. Russell, Laramie, Wyo	544
Fort Ellis, Gallatin, Mont	519
Fort Fetterman, Albany, Wyo	712
Fort Garland, Costilla, Colo	861
Fort Kearney, Kearney, Nebr	872
Fort Laramie, Laramie, Wyo	622
Fort Leavenworth, Leavenworth, Kans	1,251
Fort McPherson, Lincoln, Nebr	787
Fort Sanders, (Laramie City,) Albany, Wyo	498
Fort Sedgewick, Wild, Colo	684
Fort Fred Steele, Carbon, Wyo	365
North Platte, Lincoln, Nebr	757
Ogden City, Weber, Utah	39
Omaha City, Douglas, Nebr	1,068

CAMP DOUGLAS, UTAH, TO—

	Miles.
San Francisco, San Francisco, Cal	865
Salt Lake City, Salt Lake, Utah	3
Washington, D. C	2,337
Virginia City, Madison, Mont	441

CAMP GASTON, CAL., TO—

Arcata, Humboldt, Cal	40
Camp Grant, Pima, Ariz	1,339
Camp Goodwin, Pima, Ariz	1,492
Camp Halleck, Elko, Nev	769
Camp Hualpai, Yavapai, Nev	1,303
Camp Independence, Inyo, Cal	712
Camp McDermitt, Humboldt, Nev	675
Camp Mohave, Mohave, Ariz	1,093
Camp Verde, Yavapai, Ariz	1,303
Camp Wright, Mendocino, Cal	265
Carson, Ormsby, Nev	463
Cloverdale, Sonoma, Cal	233
Covelo, Mendocino, Cal	264
Ehrenberg, Yuma, Ariz	1,053
Eureka, Humboldt, Cal	40
Fort Hall, Oneida, Idaho	1,130
Fort Klamath, Wasco, Oreg	337
Fort Whipple, (Prescott,) Yavapai, Ariz	1,264
Fort Yuma, San Diego, Cal	1,048
Healdsburgh, Sonoma, Cal	251
Jacksonville, Jackson, Oreg	237
Junction Station, Placer, Cal	289
Los Angeles, Los Angeles, Cal	781
Maricopa Wells, Pima, Ariz	1,248
Marysville, Yuba, Cal	255
Napa Junction, Napa, Cal	336
Orleans, Klamath, Cal	33
Prescott, Yavapai, Ariz	1,264
Reno, Washoe, Nev	425
Sacramento, Sacramento, Cal	307
San Bernardino, San Bernardino, Cal	843
San Diego, San Diego, Cal	851
San Francisco, San Francisco, Cal	334
Tehama, Tehama, Cal	179
Tucson, Pima, Ariz	1,373
Ukiah, Mendocino, Cal	202
Vallego, Solano, Cal	343
Washington, D. C	3,311
Wickenburgh, Yavapai, Ariz	1,173
Wilmington, Los Angeles, Cal	761

CAMP GRANT, ARIZ., (OLD,) TO—

Camp Grant, Ariz., (new,)	140
Camp Halleck, Elko, Nev	1 376
Camp Hualpai, Yavapai, Nev	312
Camp Independence, Inyo, Cal	881
Camp Lincoln, ———, Cal	1,398

CAMP GRANT, ARIZ., (OLD,) TO—

	Miles.
Camp McDermitt, Humboldt, Nev	1,401
Camp McDowell, Maricopa, Ariz	105
Camp Mohave, Mohave, Ariz	439
Camp Pinal, Maricopa, Ariz	60
Camp Reno, Maricopa, Ariz	138
Camp Verde, Yavapai, Ariz	316
Camp Wallen, Pima, Ariz	173
Camp Wright, Mendocino, Cal	1,200
Ehrenberg, Yuma, Ariz	306
Elko, Elko, Nev	1,346
Florence, Pima, Ariz	55
Fort Bliss, El Paso, Tex	413
Fort Whipple, Yavapai, Ariz	277
Fort Yuma, San Diego, Cal	291
Hardyville, Mohave, Ariz	446
Los Angeles, Los Angeles, Cal	578
Maricopa Wells, Pima, Ariz	91
Mesilla, Doña Aña, N. Mex	360
Prescott, Yavapai, Ariz	277
Sacramento, Sacramento, Cal	997
San Bernardino, San Bernardino, Cal	516
San Diego, San Diego, Cal	488
San Francisco, San Francisco, Cal	1,005
Tucson, Pima, Ariz	60
Washington, D. C	2,788
Wickenburgh, Yavapai, Ariz	186
Wilmington, Los Angeles, Cal	598

CAMP HALLECK, NEV., TO—

Camp Harney, Grant, Oreg	439
Camp Hualpai, Yavapai, Nev	1,063
Camp Independence, Inyo, Cal	631
Camp Lincoln, ———, Cal	879
Camp Lowell, (new,) Pima, Ariz	1,416
Camp McDermitt, Humboldt, Nev	254
Camp McDowell, Pima, Ariz	1,299
Camp Mohave, Mohave, Ariz	937
Camp Three Forks, Owyhee, Idaho	380
Camp Warner, Grant, Oreg	647
Camp Wright, Mendocino, Cal	776
Canyon City, Grant, Oreg	654
Cheyenne, Laramie, Wyo	821
Corinne, Box Elder, Utah	280
Dalles (The), Wasco, Oreg	834
Denver, Arapahoe, Colo	927
Elko, Elko, Nev	30
Fort Boisé, Ada, Idaho	454
Fort Colville, Stevens, Wash	978
Fort Hall, Oneida, Idaho	421
Fort Klamath, Wasco, Oreg	437
Fort Lapwai, Nez Percés, Idaho	822
Fort Leavenworth, Leavenworth, Kans	1,517
Fort Stevens, Clatsop, Oreg	1,058
Fort Whipple, (Prescott,) Yavapai, Ariz	1,162
Fort Yuma, San Diego, Cal	1,295
Halleck Station, Elko, Nev	12
Kelton, Box Elder, Utah	214
Los Angeles, Los Angeles, Cal	1,028

CAMP HALLECK, NEV., TO—

	Miles.
Ogden City, Weber, Utah	305
Omaha, Douglas, Nebr	1,334
Portland, Multnomah, Oreg	954
Prescott, Yavapai, Ariz	1,162
Reno, Washoe, Nev	344
Sacramento, Sacramento, Cal	498
San Bernardino, San Bernardino, Cal	1,090
San Diego, San Diego, Cal	1,098
San Francisco, San Francisco, Cal	581
San Juan Island, Whatcom, Wash	1,116
Sitka, Alaska	1,939
Tucson, Pima, Ariz	1,410
Umatilla, Umatilla, Oreg	728
Vancouver, Clarke, Wash	836
Victoria, Brit. Col	1,136
Walla Walla, Walla Walla, Wash	675
Wallula, Walla Walla, Wash	703
Washington, D. C	2,603
Wilmington, Los Angeles, Cal	1,008
Winnemucca, Humboldt, Nev	174

CAMP HARNEY, OREG., TO—

Camp Independence, Inyo, Cal	722
Camp McDermitt, Humboldt, Nev	185
Camp Three Forks, Owyhee, Idaho	220
Camp Warner, Grant, Oreg	130
Camp Wright, Mendocino, Cal	1,011
Canyon City, Grant, Oreg	75
Cape Disappointment, Pacific, Wash	486
Cheyenne City, Laramie, Wyo	1,197
Corinne, Box Elder, Utah	659
Elko, Elko, Nev	409
Fort Boisé, Ada, Idaho	225
Fort Colville, Stevens, Wash	663
Fort Hall, Oneida, Idaho	732
Fort Klamath, Wasco, Oreg	260
Fort Lapwai, Nez Percés, Idaho	507
Fort Leavenworth, Leavenworth, Kans	1,896
Fort Stevens, Clatsop, Oreg	479
Kelton, Box Elder, Utah	593
Ogden City, Weber, Utah	684
Omaha City, Douglas, Nebr	1,713
Portland, Multnomah, Oreg	375
Reno, Washoe, Nev	435
Sacramento, Sacramento, Cal	589
San Francisco, San Francisco, Cal	672
San Juan Island, Whatcom, Wash	677
The Dalles, Wasco, Oreg	255
Umatilla, Umatilla, Oreg	355
Vancouver, Clarke, Wash	357
Victoria, Brit. Col	707
Walla Walla, Walla Walla, Wash	405
Wallula, Walla Walla, Wash	380
Washington, D. C	2,982
Winnemucca, Humboldt, Nev	265

CAMP HUALPAI, ARIZ., TO—

	Miles.
Camp Independence, Inyo, Cal	885
Camp Lowell, (new,) Pima, Ariz	352
Camp McDermitt, Humboldt, Nev	1,357
Camp McDowell, Pima, Ariz	235
Camp Mohave, Mohave, Ariz	126
Camp Verde, Yavapai, Ariz	78
Camp Wright, Mendocino, Cal	1,080
Ehrenberg, Yuma, Ariz	246
Elko, Elko, Nev	1,033
Fort Whipple, Yavapai, Ariz	39
Fort Yuma, San Diego, Cal	353
Los Angeles, Los Angeles, Cal	438
Maricopa Wells, Pima, Ariz	221
Ogden City, Weber, Utah	1,033
Prescott, Yavapai, Ariz	39
San Bernardino, San Bernardino, Cal	376
San Diego, San Diego, Cal	550
San Francisco, San Francisco, Cal	885
Tucson, Pima, Ariz	346
Washington, D. C	3,331
Wickenburgh, Yavapai, Ariz	126
Wilmington, Los Angeles, Cal	458

CAMP INDEPENDENCE, CAL., TO—

Aurora, Esmeralda, Nev	133
Camp Lincoln, ———, Cal	822
Camp Lowell, (new,) Pima, Ariz	921
Camp McDermitt, Humboldt, Nev	537
Camp McDowell, Pima, Ariz	804
Camp Mohave, Mohave, Ariz	759
Camp Reno, Pima, Ariz	837
Camp Verde, Yavapai, Ariz	825
Camp Wallen, Pima, Ariz	956
Camp Wright, Mendocino, Cal	719
Carson City, Ormsley, Nev	249
Ehrenberg, Yuma, Ariz	575
Fort Hall, Oneida, Idaho	992
Fort Leavenworth, Leavenworth, Kans	2,088
Fort Whipple, Yavapai, Ariz	786
Fort Yuma, San Diego, Cal	644
Havilah, Kern, Cal	153
Los Angeles, Los Angeles, Cal	303
Maricopa Wells, Pima, Ariz	790
Olympia, Thurston, Wash	1,161
Omaha City, Douglas, Nebr	1,905
Portland, Multnomah, Oreg	1,023
Prescott, Yavapai, Ariz	786
Reno, Washoe, Nev	287
Sacramento, Sacramento, Cal	441
Sacramento Junction, Sacramento, Cal	423
San Bernardino, San Bernardino, Cal	365
San Diego, San Diego, Cal	447
San Francisco, San Francisco, Cal	524
San Juan Island, Whatcom, Wash	1,283
Sitka, Alaska	2,098

CAMP INDEPENDENCE, CAL., TO—

	Miles.
Tucson, Pima, Ariz	915
Victoria, Brit. Amer	1,303
Visalia, Tulare, Cal	268
Washington, D. C	3,173
Wickenburgh, Yavapai, Ariz	695
Wilmington, Los Angeles, Cal	323

CAMP LOWELL, TUCSON, ARIZ., (NEW,) TO—

Tucson, Pima, Ariz	6
(This port is six miles easterly from Tucson, and all distances from the west will be calculated *via* Tucson.)	

CAMP LOWELL, ARIZ., (OLD,)

(Was located at Tucson.)

CAMP McDERMITT, NEV., TO—

Camp McDowell, Maricopa, Ariz	1,324
Camp Mohave, Mohave, Ariz	1,131
Camp Reno, Pima, Ariz	1,357
Camp Verde, Yavapai, Ariz	1,341
Camp Three Forks, Owyhee, Idaho	168
Camp Warner, Grant, Oreg	629
Camp Wright, Mendocino, Cal	678
Canyon City, Grant, Oreg	400
Cape Disappointment, Pacific, Oreg	811
Cheyenne, Laramie, Wyo	1,011
Dalles, (The,) Wasco, Oreg	580
Ehrenberg, Yuma, Ariz	1,206
Elko, Elko, Nev	224
Fort Boisé, Ada, Idaho	200
Fort Colville, Stevens, Wash	724
Fort D. A. Russell, Laramie, Wyo	1,013
Fort Hall, Oneida, Idaho	611
Fort Klamath, Wasco, Oreg	643
Fort Lapwai, Nez Percés, Idaho	568
Fort Leavenworth, Leavenworth, Kans	1,711
Fort Stevens, Clatsop, Oreg	804
Fort Whipple, Yavapai, Ariz	1,306
Fort Yuma, San Diego, Cal	1,201
Lewiston, Nez Percés, Idaho	556
Los Angeles, Los Angeles, Cal	823
Maricopa Wells, Pima, Ariz	1,401
Ogden City, Weber, Utah	499
Omaha City, Douglas, Nebr	1,528
Portland, Multnomah, Oreg	700
Prescott, Yavapai, Ariz	1,306
Reno, Washoe, Nev	250

CAMP McDERMITT, NEV., TO—

	Miles.
Sacramento, Sacramento, Cal	404
San Bernardino, San Bernardino, Cal	885
San Diego, San Diego, Cal	1,004
San Francisco, San Francisco, Cal	487
Tucson, Pima, Ariz	1,435
Washington, D. C	2,797
Wickenburgh, Yavapai, Ariz	1,215
Wilmington, Los Angeles, Cal	843
Winnemucca, Humboldt, Nev	80

CAMP McDOWELL, ARIZ., TO—

Camp Mohave, Mohave, Ariz	362
Camp Reno, Pima, Ariz	33
Camp Verde, Yavapai, Ariz	239
Camp Wallen, Pima, Ariz	186
Camp Wright, Mendocino, Cal	1,143
Ehrenberg, Yuma, Ariz	229
Elko, Elko, Nev	1,269
Fort Bliss, El Paso, Tex	476
Fort Cummings, Grant, N. Mex	365
Fort Yuma, San Diego, Cal	264
Los Angeles, Los Angeles, Cal	501
Maricopa Wells, Pima, Ariz	64
Mesilla, Doña Aña, N. Mex	425
Ogden City, Weber, Utah	994
Phœnix, Yavapai, Ariz	39
Prescott, Yavapai, Ariz	200
Sacramento, Sacramento, Cal	1,031
San Bernardino, San Bernardino, Cal	439
San Diego, San Diego, Cal	461
San Francisco, San Francisco, Cal	948
Santa Fé, Santa Fé, N. Mex	735
Tucson, Pima, Ariz	125
Washington, D. C	2,853
Wickenburgh, Yavapai, Ariz	109
Wilmington, Los Angeles, Cal	521

CAMP MOHAVE, ARIZ., TO—

Camp Reno, Pima, Ariz	395
Camp Verde, Yavapai, Ariz	264
Camp Wallen, Pima, Ariz	534
Camp Wright, Mendocino, Cal	954
Ehrenberg, Yuma, Ariz	140
Elko, Elko, Nev	907
Fort Bliss, El Paso, Tex	826
Fort Cummings, Grant, N. Mex	713
Fort Whipple, Yavapai, Ariz	225
Fort Yuma, San Diego, Cal	240
Hardyville, Mohave, Ariz	7
Los Angeles, Los Angeles, Cal	312
Maricopa Wells, Pima, Ariz	348

CAMP MOHAVE, ARIZ., TO—

	Miles.
Ogden City, Weber, Utah	632
Phœnix, Yavapai, Ariz	323
Prescott, Yavapai, Ariz	225
Sacramento, Sacramento, Cal	731
Saint George, Washington, Utah	247
Salt Lake City, Salt Lake, Utah	595
San Bernardino, San Bernardino, Cal	250
San Diego, San Diego, Cal	437
San Francisco, San Francisco, Cal	759
Tucson, Pima, Ariz	473
Washington, D. C	2,930
Wickenburgh, Yavapai, Ariz	253
Wilmington, Los Angeles, Cal	332

CAMP STAMBAUGH, WYO., TO—

Atlantic City, Sweet Water, Wyo	2½
Bozeman, Gallatin, Mont	758
Bryan Station, Uintah, Wyo	110
Cheyenne City, Laramie, Wyo	452
Corinne, Box Elder, Utah	305
Fort Bridger, Uintah, Wyo	165
Fort D. A. Russell, Laramie, Wyo	454
Fort Ellis, Gallatin, Mont	761
Fort Fetterman, Albany, Wyo	622
Fort Fred Steele, Carbon, Wyo	274
Fort Garland, Costilla, Colo	771
Fort Hall, Oneida, Idaho	445
Fort Kearney, Kearney, Nebr	882
Fort Laramie, Laramie, Wyo	532
Fort Leavenworth, Leavenworth, Kans	1,493
Fort McPherson, Lincoln, Nebr	697
Fort Sanders, Albany, Wyo	398
Fort Sedgwick, Wild, Colo	594
Helena, Lewis & Clarke, Mont	750
North Platte, Lincoln, Nebr	697
Ogden City, Weber, Utah	281
Omaha City, Douglas, Nebr	948
San Francisco, San Francisco, Cal	1,107
Salt Lake City, Salt Lake, Utah	318
Washington, D. C	2,234
Virginia City, Madison, Mont	683

CAMP SUPPLY, IND. TER., TO—

Fort Dodge, Ford, Kans	91
Fort Harker, Ellsworth, Kans	225
Fort Hays, Ellis, Kans	171
Fort Larned, Pawnee, Kans	152
Fort Leavenworth, Leavenworth, Kans	461
Washington, D. C	1,550

CAMP THREE FORKS, IDAHO, TO—

	Miles.
Camp Warner, Grant, Oreg	240
Camp Wright, Mendocino, Cal	847
Canyon City, Grant, Oreg	305
Cape Disappointment, Pacific, Wash	716
Elko, Elko, Nev	389
Fort Boisé, Ada, Idaho	105
Fort Colville, Stevens, Wash	629
Fort Hall, Oneida, Idaho	562
Fort Klamath, Wasco, Oreg	370
Fort Lapwai, Nez Percés, Idaho	473
Fort Leavenworth, Leavenworth, Kans	1,659
Fort Stevens, Clatsop, Oreg	709
Fort Tongass, Alaska	1,497
Fort Wrangel, Alaska	1,675
Kelton, Box Elder, Utah	355
Lewiston, Nez Percés, Idaho	461
Ogden City, Weber, Utah	447
Olympia, Thurston, Wash	745
Omaha City, Douglas, Nebr	1,476
Portland, Multnomah, Oreg	605
Port Townsend, Jefferson, Wash	877
Reno, Washoe, Nev	415
Sacramento, Sacramento, Cal	569
San Francisco, San Francisco, Cal	652
San Juan Island, Whatcom, Wash	907
Sitka, Alaska	1,680
The Dalles, Wasco, Oreg	485
Umatilla, Umatilla, Oreg	431
Vancouver, Clarke, Wash	587
Victoria, Brit. Amer	927
Walla Walla, Walla Walla, Wash	376
Wallula, Walla Walla, Wash	406
Washington, D. C	2,745
Winnemucca, Humboldt, Nev	245

CAMP VERDE, ARIZ., TO—

Camp Wallen, Pima, Ariz	411
Camp Wright, Mendocino, Cal	1,154
Ehrenberg, Yuma, Ariz	250
Fort Bliss, El Paso, Tex	703
Fort Cummings, Grant, N. Mex	590
Fort Whipple, Yavapai, Ariz	39
Fort Yuma, San Diego, Cal	445
Hardyville, Mohave, Ariz	271
Los Angeles, Los Angeles, Cal	522
Maricopa Wells, Pima, Ariz	225
Mesilla, Doña Aña, N. Mex	650
Ogden City, Weber, Utah	896
Phœnix, Yavapai, Ariz	200
Prescott, Yavapai, Ariz	39
Sacramento, Sacramento, Cal	1,042
Saint George, Washington, Utah	511
Salt Lake City, Salt Lake, Utah	859
San Bernardino, San Bernardino, Cal	460
San Diego, San Diego, Cal	622

CAMP VERDE, ARIZ., TO—

	Miles.
San Francisco, San Francisco, Cal	959
Tucson, Pima, Ariz	350
Washington, D. C	3,194
Wickenburgh, Yavapai, Ariz	130
Wilmington, Los Angeles, Cal	542

CAMP WARNER, OREG., TO—

Camp Wright, Mendocino, Cal	735
Canyon City, Grant, Oreg	205
Cape Disappointment, Pacific, Wash	616
Elko, Elko, Nev	617
Fort Boisé, Ada, Idaho	405
Fort Colville, Stevens, Wash	793
Fort Hall, Oneida, Idaho	862
Fort Klamath, Wasco, Oreg	130
Fort Lapwai, Nez Percés, Idaho	637
Fort Leavenworth, Leavenworth, Kans	2,105
Fort Stevens, Clatsop, Oreg	609
Fort Tongass, Alaska	1,397
Fort Wrangel, Alaska	1,575
Kelton, Box Elder, Utah	801
Lewiston, Nez Percés, Idaho	625
Ogden City, Weber, Utah	892
Omaha City, Douglas, Nebr	1,921
Portland, Multnomah, Oreg	505
Reno, Washoe, Nev	303
Sacramento, Sacramento, Cal	457
San Francisco, San Francisco, Cal	445
San Juan Island, Whatcom, Wash	807
Sitka, Alaska	1,580
The Dalles, Wasco, Oreg	385
Umatilla, Umatilla, Oreg	485
Vancouver, Clarke, Wash	487
Walla Walla, Walla Walla, Wash	538
Wallula, Walla Walla, Wash	510
Washington, D. C	3,489
Winnemucca, Humboldt, Nebr	473

CAMP WRIGHT, CAL., TO—

Cloverdale, Sonoma, Cal	94
Covelo, Mendocino, Cal	1
Ehrenberg, Yuma, Ariz	914
Eureka, Humboldt, Cal	225
Fort Whipple, Yavapai, Ariz	1,125
Fort Yuma, San Diego, Cal	909
Healdsburgh, Sonoma, Cal	112
Los Angeles, Los Angeles, Cal	642
Maricopa Wells, Pima, Ariz	1,109
Petaluma, Sonoma, Cal	144
Prescott, Yavapai, Ariz	1,125
San Bernardino, San Bernardino, Cal	704
San Diego, San Diego, Cal	712
San Francisco, San Francisco, Cal	195
Washington, D. C	3,318
Wickenburgh, Yavapai, Ariz	1,034
Wilmington, Los Angeles, Cal	622

CANYON CITY, OREG., TO—

	Miles.
Cape Disappointment, Pacific, Wash	411
Corinne, Box Elder, Utah	457
Elko, Elko, Nev	624
Fort Boisé, Ada, Idaho	200
Fort Colville, Stevens, Wash	588
Fort Hall, Oneida, Idaho	657
Fort Klamath, Wasco, Oreg	335
Fort Lapwai, Nez Percés, Idaho	436
Fort Leavenworth, Leavenworth, Kans	1,554
Fort Stevens, Clatsop, Oreg	404
Fort Tongass, Alaska	1,192
Fort Wrangel, Alaska	1,370
Kelton, Box Elder, Utah	450
Ogden City, Weber, Utah	542
Olympia, Thurston, Wash	440
Omaha City, Douglas, Nebr	1,569
Port Townsend, Jefferson, Wash	572
Reno, Washoe, Nev	650
Sacramento, Sacramento, Cal	804
San Francisco, San Francisco, Cal	887
San Juan Island, Whatcom, Wash	602
Sitka, Alaska	1,375
The Dalles, Wasco, Oreg	180
Umatilla, Umatilla, Oreg	280
Vancouver, Clarke, Wash	282
Walla Walla, Walla Walla, Wash	335
Wallula, Walla Walla, Wash	305
Washington, D. C	2,840
Winnemucca, Humboldt, Nev	480

CAPE DISAPPOINTMENT, PACIFIC, WASH., TO—

Eugene City, Lane, Oreg	225
Fort Boisé, Ada, Idaho	611
Fort Colville, Stevens, Wash	637
Fort Hall, Oneida, Idaho	1,068
Fort Klamath, Wasco, Oreg	521
Fort Lapwai, Nez Percés, Idaho	483
Fort Stevens, Clatsop, Oreg	8
Jacksonville, Jackson, Oreg	404
Lewiston, Nez Percés, Idaho	471
Monticello, Cowlitz, Wash	71
Olympia, Thurston, Wash	156
Oysterville, Pacific, Wash	25
Portland, Multnomah, Oreg	111
San Francisco, San Francisco, Cal	657
Sitka, Alaska	960
The Dalles, Wasco, Oreg	231
Umatilla, Umatilla, Oreg	331
Vancouver, Clarke, Wash	129
Victoria, Brit. Amer	298
Walla Walla, Walla Walla, Wash	384
Wallula, Walla Walla, Wash	356
Washington, D. C	3,750

CEDAR KEYS, FLA., TO—

	Miles.
Fort Barrancas, Escambia, Fla	326
Fort Jefferson, Tortugas, Fla	394
Fernandina, Nassau, Fla	154
Key West, Monroe, Fla	330
New Orleans, Orleans, La	558
Pensacola, Escambia, Fla	318
Savannah, Chatham, Ga	266
Saint Mark's, Wakulla, Fla	96
Washington, D. C	964

CHATTANOOGA, TENN., TO—

Crab Orchard, Lincoln, Ky	297
Fort Barrancas, Escambia, Fla	447
Fort Jefferson, Tortugas, Fla	1,214
Fort Pulaski, Chatham, Ga	456
Frankfort, Franklin, Ky	366
Humboldt, Gibson, Tenn	291
Huntsville, Madison, Ala	98
Jackson, Hinds, Miss	468
Key West, Monroe, Fla	1,150
Lebanon, Marion, Ky	216
Louisville, Jefferson, Ky	336
Meridian, Lauderdale, Miss	295
Mobile, Mobile, Ala	462
Mount Sterling, Montgomery, Ky	391
Nashville, Davidson, Tenn	151
Paducah, McCracken, Ky	395
Pensacola, Escambia, Fla	439
Shelbyville, Shelby, Ky	367
Saint Augustine, St. John's, Fla	649
Washington, D. C	626

CHAMPLAIN ARSENAL, VERGENNES, VT., TO—

Albany, Albany, N. Y	127
Baltimore, Baltimore, Md	460
Boston, Suffolk, Mass	213
Burlington, Chittenden, Vt	21
Chatham Four Corners, Columbia, N. Y	154
Fort Adams, Newport, R. I	268
Fort Columbus, New York, N, Y	271
Fort Foote, Prince George's, Md	508
Fort Hamilton, Kings, N. Y	278
Fort Indpendence, Suffolk, Mass	215
Fort Johnson, Brunswick, N. C	909
Fort Macon, Carteret, N. C	879
Fort McHenry, Baltimore, Md	463
Fort Monroe, Elizabeth City, Va	645
Fort Preble, Cumberland, Me	240
Fort Sullivan, Washington, Me	500
Fort Trumbull, New London, Conn	245
Fort Wadsworth, Richmond, N. Y	279

CHAMPLAIN ARSENAL, VERGENNES, VT., TO—

	Miles.
Fort Warren, Suffolk, Mass	220
Fort Washington, Prince George's, Md	513
Fort Wood, New York, N. Y	272
Frankford Arsenal, Philadelphia, Pa	351
Newport, Newport, R. I	267
New York, New York, N. Y	270
Philadelphia, Philadelphia, Pa	359
Pikesville Arsenal, Baltimore, Md	468
Plattsburgh Barracks, Clinton, N. Y	43
Raleigh, Wake, N. C	814
Rutherfordton, Rutherford, N. C	992
Rutland, Rutland, Vt	46
Washington, D. C	498
Watertown Arsenal, Middlesex, Mass	221
Willett's Point, Queens, N. Y	290
Wilmington, New Hanover, N. C	879

CHEYENNE AGENCY, DAK., TO—

Bismarck, Dak	245
Crow Creek Agency, Buffalo, Dak	128
Fort Abercrombie, Shyenne, Dak	477
Fort Abraham Lincoln, Dak	245
Fort Benton, Choteau, Mont	1,220
Fort Buford, Howard, Dak	490
Fort Randall, Todd, Dak	228
Fort Rice, Morton, Dak	215
Fort Seward, Dak	345
Fort Snelling, Hennepin, Minn	882
Fort Stevenson, Steven, Dak	340
Fort Sully, Buffalo, Dak	580
Fort Thompson, Buffalo, Dak	128
Fort Totten, Ramsey, Dak	466
Moorhead City, Minn	443
Saint Paul, Ramsey, Minn	874
Sioux City, Woodbury, Iowa	359
Vermillion, Clay, Dak	326
Washington, D. C	1,674
Whetstone Agency, Gregory, Dak	214
White Swan, Charles Mix, Dak	228
Yankton, Yankton, Dak	299

CHEYENNE AGENCY, IND. TER., TO—

Fort Harker, Ellsworth, Kans	272

CHEYENNE CITY, WYO., TO—

Colorado City, El Paso, Colo	178
Corinne, Box Elder, Utah	537
Denver, Arapahoe, Colo	102
Elko, Elko, Nev	788
Fort Bridger, Uintah, Wyo	396
Fort D. A. Russell, Laramie, Wyo	2

CHEYENNE CITY, WYO., TO—

	Miles.
Fort Fetterman, Albany, Wyo	170
Fort Fred Steele, Carbon, Wyo	181
Fort Garland, Costilla, Colo	319
Fort Hall, Oneida, Idaho	677
Fort Harker, Ellsworth, Kans	522
Fort Hays, Ellis, Kans	452
Fort Kearney, Kearney, Nebr	330
Fort Laramie, Laramie, Wyo	80
Fort Larned, Pawnee, Kans	502
Fort Leavenworth, Leavenworth, Kans	699
Fort Lyon, Bent, Colo	303
Fort McPherson, Lincoln, Nebr	245
Fort Reynolds, Pueblo, Colo	244
Fort Sanders, Albany, Wyo	54
Fort Sedgwick, Weld, Colo	142
Fort Union, Mora, N. Mex	489
Fort Wallace, Wallace, Kans	318
Julesburgh, Weld, Colo	139
Kansas City, Jackson, Mo	720
Kit Carson, Greenwood, Colo	253
Laramie City, Albany, Wyo	56
Lawrence, Douglas, Kans	702
Leavenworth City, Leavenworth, Kans	694
North Platte, Lincoln, Nebr	225
Ogden City, Weber, Utah	513
Omaha City, Douglas, Nebr	516
Pueblo, Pueblo, Colo	224
Salt Lake City, Salt Lake, Utah	550
San Francisco, San Francisco, Cal	1,339
Santa Fé, Santa Fé, N. Mex	505
Trinidad, Las Animas, Colo	321
Washington, D. C	1,782

CHUGWATER STATION, WYOMING, TO—

Fort D. A. Russell, Laramie, Wyo	85
Fort Laramie, Laramie, Wyo	18
Washington, Washington, D. C	1,869

CLARKSVILLE, TEX., TO—

Denton, Denton, Tex	144
Fort Richardson, Jack, Tex	204
Hallsville, Harrison, Tex	117
Jefferson, Marion, Tex	87
Little Rock, Pulaski, Ark	239
Longview, Upshur, Tex	127
Marshall, Harrison, Tex	103
San Antonio, Socorro, Tex	392
Sherman, Grayson, Tex	101
Shreveport, Caddo, La	145
Washington, Hempstead, Ark	109
Washington, D. C	1,311

COLORADO CITY, COLO., TO—

	Miles.
Badito, Huerfano, Colo	60
Denver, Arapahoe, Colo	76
Fort Bayard, Grant, N. Mex	832
Fort Bridger, Uintah, Wyo	574
Fort Bliss, El Paso, Tex	780
Fort Craig, Socorro, N. Mex	606
Fort Cummings, Grant, N. Mex	787
Fort Garland, Costilla, Colo	141
Fort McRae, Socorro, N. Mex	638
Fort Reynolds, Pueblo, Colo	66
Fort Selden, Doña Aña, N. Mex	706
Fort Stanton, Socorro, N. Mex	658
Fort Sumner. San Miguel, N. Mex	459
Fort Union, Mora, N. Mex	311
Fort Wingate, Santa Aña, N. Mex	627
Greenhorn, Huerfano, Colo	28
Kit Carson, Greenwood, Colo	227
Pueblo, Pueblo, Colo	46
Santa Fé, Santa Fé, N. Mex	417
Trinidad, Las Animas, Colo	143
Washington, D. C	1,905

COLUMBIA, TEX., TO—

Columbus, Colorado, Tex	120
Corsicana, Navarro, Tex	261
Denison, Texas	386
Fort Bliss, El Paso, Tex	961
Fort Clark, Kinney, Tex	386
Fort Concho, Bexar, Tex	489
Fort Davis, Presidio, Tex	739
Fort Duncan, Maverick, Tex	431
Fort Gibson, Cherokee N., Ind. Ter	559
Fort Griffin, Shackelford, Tex	634
Fort McIntosh, Webb, Tex	425
Fort McKavett, Menard, Tex	435
Fort Quitman, El Paso, Tex	879
Galveston, Galveston, Tex	95
Houston, Harris, Tex	50
Matagorda, Matagorda, Tex	55
New Orleans, Orleans, La	397
Orange, Orange, Tex	160
San Antonio, Bexar, Tex	260
Victoria, Victoria, Tex	79
Waco, McLennan, Tex	241
Washington, D. C	1,593

COLUMBUS, TEX., TO—

Denison, Texas	354
Fort Bliss, El Paso, Tex	841
Fort Clark, Kinney, Tex	266
Fort Concho, Bexar, Tex	369

COLUMBUS, TEX., TO—

	Miles.
Fort Davis, Presidio, Tex	619
Fort Duncan, Maverick, Tex	311
Fort Gibson, Cherokee N., Ind. Ter	513
Fort Griffin, Shackelford, Tex	514
Fort McIntosh, Webb, Tex	305
Fort McKavett, Menard, Tex	315
Fort Quitman, El Paso, Tex	759
Fort Richardson, Jack, Tex	366
Fort Smith Sebastian, Ark	557
Galveston, Galveston, Tex	129
Hallsville, Harrison, Tex	34
Harrisburgh, Harris, Tex	84
Hillsborough, Hill, Tex	244
Houston, Harris, Tex	84
Indianola, Calhoun, Tex	125
New Orleans, Orleans, La	431
Orange, Orange, Tex	194
Port Lavaca, Calhoun, Tex	119
San Antonio, Bexar, Tex	140
Texana, Jackson, Tex	89
Victoria, Victoria, Tex	79
Waco, McLennan, Tex	209
Washington, D. C	1,627
Weatherford, Parker, Tex	325

CORINNE, UTAH, TO—

Diamond City, Meagher, Mont	485
Elko, Elko, Nev	251
Fort Benton, Chouteau, Mont	589
Fort Boisé, Ada, Idaho	317
Fort Colville, Stevens, Wash	841
Fort Ellis, Gallatin, Mont	456
Fort Hall, Oneida, Idaho	140
Fort Leavenworth, Leavenworth, Kans	1,220
Fort Shaw, Lewis and Clarke, Mont	525
Helena, Lewis and Clarke, Mont	445
Kelton, Box Elder, Utah	67
Ogden City, Weber, Utah	24
Omaha City, Douglas, Nebr	1,037
Reno, Washoe, Nev	565
San Francisco, San Francisco, Cal	802
Salt Lake City, Salt Lake, Utah	61
Spokan Bridge, Stevens, Wash	743
The Dalles, Wasco, Oreg	687
Virginia City, Madison, Mont	378
Washington, D. C	2,322
Winnemucca, Humboldt, Nev	395

CORPUS CHRISTI, TEX., TO—

Denison, Texas	650
Fort Bliss, El Paso, Tex	863
Fort Clark, Kinney, Tex	288

CORPUS CHRISTI, TEX., TO—

	Miles.
Fort Concho, Bexar, Tex	391
Fort Davis, Presidio, Tex	641
Fort Duncan, Marverick, Tex	333
Fort Gibson, Cherokee N., Ind. Ter	823
Fort Griffin, Shackelford, Tex	536
Fort McIntosh, Webb, Tex	160
Fort McKavett, Menard, Tex	337
Fort Quitman, El Paso, Tex	781
Fort Sill, Choctaw N., Ind Ter	846
Fort Stockton, Pecos, Tex	561
Galveston, Galveston, Tex	263
Goliad, Goliad, Tex	154
Houston, Harris, Tex	314
Indianola, Calhoun, Tex	133
New Orleans, Orleans, La	565
Refugio, Refugio, Tex	121
Ringgold Barracks, Starr, Tex	141
Rio Grande City, Starr, Tex	140
Rockport, Refugio, Tex	80
San Antonio, Bexar, Tex	162
Saint Mary's, Refugio, Tex	105
Victoria, Victoria, Tex	179
Waco, McLennan, Tex	505
Washington, D. C	1,761

CORSICANNA, TEX., TO—

Dallas, Dallas, Tex	61
Fort Bliss, El Paso, Tex	935
Fort Clark, Kinney, Tex	360
Fort Concho, Bexar, Tex	463
Fort Davis, Presidio, Tex	710
Fort Duncan, Maverick, Tex	405
Fort Gibson, Cherokee N., Ind. Ter	298
Fort Griffin, Shackelford, Tex	608
Fort Leavenworth, Leavenworth, Kans	556
Fort McIntosh, Webb, Tex	399
Fort McKavett, Menard, Tex	409
Fort Sill, Choctaw N., Ind. Ter	321
Fort Smith, Sebastian, Ark	325
Fort Stockton, Pecos, Tex	633
Galveston, Galveston, Tex	262
Houston, Harris, Tex	211
Indianola, Calhoun, Tex	339
Jefferson, Marion, Tex	166
New Orleans, Orleans, La	553
Ringgold Barracks, Starr, Tex	540
Rio Grande City, Starr, Tex	539
San Antonia, Bexar, Tex	234
Shreveport, Caddo, La	191
Tyler, Smith, Tex	85
Waco, McLennan, Tex	60
Washington, D. C	1,386
Waxahachie, Ellis, Tex	30
Weatherford, Parker, Tex	110

CRAB ORCHARD, KY., TO—

	Miles.
Fort Barrancas, Escambia, Fla	716
Fort Jefferson, Tortugas, Fla	1,420
Fort Pulaski, Chatham, Ga	849
Frankfort, Franklin, Ky	79
Humboldt, Gibson, Tenn	308
Huntsville, Madison, Ala	386
Jackson, Hinds, Miss	583
Key West, Monroe, Fla	1,356
Lebanon, Marion, Ky	48
Louisville, Jefferson, Ky	115
Meridian, Lauderdale, Miss	618
Mobile, Mobile, Ala	731
Mount Sterling, Montgomery, Ky	92
Nashville, Davidson, Tenn	240
Paducah, McCracken, Ky	330
Savannah, Chatham, Ga	834
Saint Augustine, St. John's, Fla	1,042
Shelbyville, Shelby, Ky	90
Washington, D. C	787

DAVID'S ISLAND, N. Y., TO—

Albany, Albany, N. Y	148
Baltimore, Baltimore, Md	212
Boston, Suffolk, Mass	218
Chatham Four Corners, Columbia, N. Y	124
Fort Adams, Newport, R. I	173
Fort Columbus, New York, N. Y	23
Fort Foote, Prince George's, Md	260
Fort Hamilton, Kings, N. Y	30
Fort Independence, Suffolk, Mass	218
Fort Johnson, Brunswick, N. C	661
Fort Macon, Carteret, N. C	644
Fort McHenry, Baltimore, Md	215
Fort Monroe, Elizabeth City, Va	397
Fort Preble, Cumberland, Me	327
Fort Sullivan, Washington, Me	587
Fort Trumbull, New London, Conn	108
Fort Wadsworth, Richmond, N. Y	31
Fort Warren, Suffolk, Mass	225
Fort Washington, Prince George's, Md	265
Fort Wood, New York, N. Y	24
Frankford Arsenal, Philadelphia, Pa	103
New Haven, New Haven, Conn	56
New Rochelle, Westchester, N. Y	2
New York, New York, N. Y	22
Philadelphia, Philadelphia, Pa	111
Pikesville Arsenal, Baltimore, Md	220
Plattsburgh Barracks, Clinton, N. Y	321
Raleigh, Wake, N. C	566
Rutherfordton, Rutherford, N. C	744
Washington, D. C	250
Watertown Arsenal, Middlesex, Mass	226
Williams' Bridge, Westchester, N. Y	8
Wilmington, New Hanover, N. C	631

DENISON, TEXAS, TO—

	Miles.
Austin, Travis, Tex	401
Columbia, Brazoria, Tex	386
Fort Bliss, El Paso, Tex	792
Fort Clark, Kinney, Tex	606
Fort Concho, Bexar, Tex	320
Fort Davis, Presidio, Tex	570
Fort Gibson, Cherokee N., Ind. Ter	173
Fort Griffin, Tex	174
Fort Leavenworth, Leavenworth, Kans	430
Fort Quitman, El Paso, Tex	710
Fort Sill, Choctaw N., Ind. Ter	196
Fort Stockton, Pecos, Tex	490
Galveston, Galveston, Tex	386
Hearne, Robertson, Tex	216
Hempstead, Austin, Tex	286
Houston, Harris, Tex	336
Kansas City, Jackson, Mo	422
Longview, Texas	391
Palestine, Anderson, Texas	307
San Antonio, Bexar, Tex	478
Washington, Washington, D. C	1,495

DENVER, COLO., TO—

Elko, Elko, Nev	890
Fort Bridger, Uintah, Wyo	498
Fort D. A. Russell, Laramie, Wyo	104
Fort Fred Steele, Carbon, Wyo	283
Fort Fettermen, Albany, Wyo	272
Fort Garland, Costilla, Colo	217
Fort Hall, Oneida, Idaho	779
Fort Harker, Ellsworth, Kans	420
Fort Kearney, Kearney, Nebr	432
Fort Laramie, Laramie, Wyo	182
Fort Leavenworth, Leavenworth, Kans	636
Fort Lyon, Bent, Colo	201
Fort McPherson, Lincoln, Nebr	347
Fort Reynolds, Pueblo, Colo	142
Fort Sedgwick, Weld, Colo	242
Fort Union, Mora, N. Mex	387
Fort Wallace, Wallace, Kans	216
Greenhorn, Huerfano, Colo	150
Julesburgh, Weld, Colo	241
Kansas City, Jackson, Mo	640
Kit Carson, Greenwood, Colo	151
Laramie City, Albany, Wyo	158
Lawrence, Douglas, Kans	600
Leavenworth City, Leavenworth, Kans	633
Omaha City, Douglas, Nebr	618
Pueblo, Pueblo, Colo	122
Salt Lake City, Salt Lake, Utah	652
San Francisco, San Francisco, Cal	1,441
Santa Fé, Santa Fé, N. Mex	493
Trinidad, Las Animas, Colo	219
Washington, D. C	1,829

DETROIT, MICH., TO—

	Miles.
Duluth, Saint Louis, Minn	745
Eagle Harbor, Keweenaw, Mich	671
Eagle River, Keweenaw, Mich	655
East Nubish Rapids, St. Mary's River	315
Fort Brady, Chippewa, Mich	339
Fort Gratiot, St. Clair, Mich	63
Fort Mackinac, Mackinac, Mich	363
Fort Wayne, Wayne, Mich	2½
Houghton, Houghton, Mich	633
Marquette, Marquette, Mich	528
Ontonagon, Ontonagon, Mich	692
Superior, Douglas, Wis	752
White Fish Point, Chippewa, Mich	381

DRUM BARRACKS, WILMINGTON, CAL., TO—

	Miles.
Ehrenberg, Yuma, Ariz	293
Fort Whipple, Yavapai, Ariz	504
Fort Yuma, San Diego, Cal	364
Los Angeles, Los Angeles, Cal	21
Maricopa Wells, Pima, Ariz	508
Prescott, Yavapai, Ariz	504
Sacramento, Sacramento, Cal	511
San Bernardino, San Bernardino, Cal	83
San Diego, San Diego, Cal	167
San Francisco, San Francisco, Cal	428
Tucson, Pima, Ariz	633
Washington, D. C	3,552
Wickenburgh, Yavapai, Ariz	413
Wilmington, Los Angeles, Cal	1

DULUTH, MINN., TO—

	Miles.
Bayfield, Bayfield, Wis	87
Fort Abercrombie, Shyenne, Dak	301
Fort Brady, (Sault de Ste. Marie,) Chippewa, Mich	560
Fort Mackinac, Mackinac, Mich	652
Ontonagon, Ontonagon, Mich	207
Red River Crossing N. Pac. R. Rd. (Moorhead,) Clay, Minn	251
St. Paul, Ramsey, Minn	156
Thomson, Carlton, Minn	23
Washington, D. C	1,371

EAGLE HARBOR, MICH., TO—

	Miles.
Copper Harbor, Keweenaw, Mich	18
Eagle River, Keweenaw, Mich	9
Fort Brady, Chippewa, Mich	332
Fort Gratiot, St. Clair, Mich	608
Fort Mackinac, Mackinac, Mich	424

EAGLE HARBOR, MICH., TO—

	Miles
Fort Wayne, Wayne, Mich	674
Houghton, Houghton, Mich	38
Marquette, Marquette, Mich	143
Ontonagon, Ontonagon, Mich	97
Superior, Douglas, Wis	297
Washington, D. C	1,294
White Fish Point Chippewa, Mich	287

EAGLE RIVER, MICH., TO—

Copper Harbor, Keweenaw, Mich	27
Fort Brady, Chippewa, Mich	323
Fort Gratiot, St. Clair, Mich	599
Fort Mackinac, Mackinac, Mich	415
Fort Wayne, Wayne, Mich	665
Houghton, Houghton, Mich	29
Marquette, Marquette, Mich	134
Ontonagon, Ontonagon, Mich	88
Superior, Douglas, Wis	288
Washington, D. C	1,285
White Fish Point, Chippewa, Mich	295

EHRENBERG, ARIZ., TO—

Fort Whipple, Yavapai, Ariz	211
Fort Yuma, San Diego, Cal	100
La Paz, Yuma, Ariz	7
Los Angeles, Los Angeles, Cal	272
Maricopa Wells, Pima, Ariz	215
Ogden City, Weber, Utah	772
Prescott, Yavapai, Ariz	211
San Bernardino, San Bernardino, Cal	210
San Diego, San Diego, Cal	297
San Francisco, San Francisco, Cal	719
Tucson, Pima, Ariz	340
Washington, D. C	3,070
Wickenburgh, Yavapai, Ariz	120
Wilmington, Los Angeles, Cal	292

ELKO, NEV., TO—

Eugene City, Lane, Oreg	936
Fort Boisé, Ada, Idaho	424
Fort Colville, Stevens, Wash	948
Fort Hall, Onedia, Idaho	391
Fort Klamath, Wasco, Oreg	707
Fort Lapevai, Nez Percés, Idaho	792
Fort Whipple, Yavapai, Ariz	1,132
Fort Yuma, San Diego, Cal	1,265
Jacksonville, Jackson, Oreg	757

ELKO, NEV., TO

	Miles.
Kelton, Box Elder, Utah	184
Los Angeles, Los Angeles, Cal	887
Ogden City, Weber, Utah	275
Olympia, Thurston, Wash	1,064
Omaha City, Douglas, Nebr	1,304
Portland, Multnomah, Oreg	924
Port Townsend, Jefferson, Wash	1,196
Prescott, Yavapai, Ariz	1,132
Red Bluff, Tehama, Cal	570
Reno, Washoe, Nev	314
Sacramento, Sacramento, Cal	468
Sacramento Junction, Sacramento, Cal	450
San Bernardino, San Bernardino, Cal	949
San Diego, San Diego, Cal	1,068
San Francisco, San Francisco, Cal	551
San Juan Island, Whatcom, Wash	1,226
Sitka, Alaska	1,999
Umatilla, Umatilla, Oreg	698
Vancouver, Clarke, Wash	906
Walla Walla, Walla Walla, Wash	695
Wallula, Walla Walla, Wash	673
Washington, D. C	2,573
Wilmington, Los Angeles, Cal	907
Winnemucca, Humboldt, Nev	144

EL PASO, TEX., TO—

Camp Bowie, Ariz	275
Doña Aña, Doña Aña, N. Mex	59
Fort Bayard, Grant, N. Mex	155
Fort Bliss, El Paso, Tex	3
Fort Bowie, Pima, Ariz	238
Fort Cummings, Grant, N. Mex	110
Fort Quitman, El Paso, Tex	85
Fort Selden, Doña Aña, N. Mex	71
Las Cruces, Doña Aña, N. Mex	53
Leasburgh, Doña Aña, N. Mex	70
Magoffinsville, El Paso, Tex	2
Mesilla, Doña Aña, N. Mex	50
Ralston, Grant, N. Mex	227
San Antonio, Bexar, Tex	704
San Elisario, El Paso, Tex	25
Santa Fé, Santa Fé, N. Mex	360
Washington, D. C	2,460

FORT ABERCROMBIE, DAK., TO—

Benson Station, Chippewa, Minn	103
Breckinridge, Wilkin, Minn	14
Duluth, St. Louis, Minn	286
Fort Benton, Choteau, Mont	1,207
Fort Buford, Howard, Dak	477
Fort Garry, Brit. Amer	257

FORT ABERCROMBIE, DAK., TO—

	Miles.
Fort Pembina, Pembina, Dak	186
Fort Randall, Todd, Dak	646
Fort Ransom, Ransom, Dak	68
Fort Rice, Morton, Dak	261
Fort Ripley, Morrison, Minn	220
Fort Seward, Dak	132
Fort Snelling, Hennepin, Minn	233
Fort Stevenson, Stevens, Dak	239
Fort Sully, Sully, Dak	518
Fort Thompson, Buffalo, Dak	593
Fort Totten, Ramsey, Dak	212
Fort Wadsworth, Deuel, Dak	76
Georgetown, Clay, Minn	52
Holy Cross, Clay, Minn	26
Lake Winnipeg, Brit. Amer	300
McCauleyville, Wilkin, Minn	1
Mendota, Dakota, Minn	235
Minneapolis, Hennepin, Minn	235
Moorhead, Clay, Minn	34
Pembina, Pembina, Dak	189
Ponka Agency, Todd, Dak	632
Saint Cloud, Fond du Lac, Minn	170
Saint Joseph, Pembina, Dak	219
Saint Paul, Ramsey, Minn	231
Sioux City, Woodbury, Iowa	503
Stone Fort, Brit. Amer	277
Vermillion, Clay, Dak	548
Washington, D. C	1,454
Whetstone Agency, Gregory, Dak	662
Yankton, Yankton, Dak	564

FORT ABRAHAM LINCOLN, DAK., TO—

Bismarck, Burleigh, Dak	4
Fort Benton, Choteau, Mont	975
Fort Buford, Howard, Dak	245
Fort Pembina, Pembina, Dak	350
Fort Randall, Todd, Dak	473
Fort Rice, Morton, Dak	25
Fort Seward, Stutsman, Dak	100
Fort Snelling, Hennepin, Minn	467
Fort Stevenson, Stevens, Dak	95
Fort Sully, Sully, Dak	283
Fort Thompson, Buffalo, Dak	368
Mendota, Dakota, Minn	469
Minneapolis, Hennepin, Minn	454
Moorhead, Clay, Minn	202
St. Paul, Ramsey, Minn	479
Sioux City, Woodbury, Iowa	599
Washington, D. C	1,667
Whetstone Agency, Gregory, Dak	459
Yankton, Yankton, Dak	539

FORT ADAMS, R. I., TO--

	Miles.
Albany, Albany, N. Y	236
Baltimore, Baltimore, Md	361
Boston, Suffolk, Mass	70
Fort Columbus, New York, N. Y	172
Fort Foote, Prince George's, Md	401
Fort Hamilton, Kings, N. Y	179
Fort Independence, Suffolk, Mass	73
Fort Johnston, Brunswick, N. C	810
Fort Macon, Carteret, N. C	793
Fort McHenry, Baltimore, Md	364
Fort Preble, Cumberland, Md	179
Fort Sullivan, (Eastport,) Washington, Me	439
Fort Trumbull, New London, Conn	100
Fort Wadsworth, Richmond, N. Y. Harb	180
Fort Warren, Suffolk, Mass	77
Fort Washington, Prince George's, Md	414
Fort Wood, New York, N. Y. Harb	173
Newport, Newport, R. I	1
New York, New York, N. Y	171
Philadelphia, Philadelphia, Pa	260
Pikesville Arsenal, Baltimore, Md	369
Plattsburgh Barracks, Clinton, N. Y	311
Raleigh, Wake, N. C	715
Rutherfordton, Rutherford, N. C	893
Washington, D. C	399
Watertown Arsenal, Middlesex, Mass	78
Willett's Point, Queens, N. Y	191
Wilmington, New Hanover, N. C	780

FORT ARBUCKLE, IND. TER., TO—

Academy, Choctaw N., Ind. Ter	53
Boggy Depot, Choctaw N., Ind. Ter	65
Fort Bliss, El Paso, Tex	902
Fort Clark, Kinney, Tex	971
Fort Concho, Bexar, Tex	430
Fort Davis, Presidio, Tex	680
Fort Duncan, Maverick, Tex	716
Fort Gibson, Choctaw N., Ind. Ter	232
Fort Griffin, Shackelford, Tex	290
Fort Leavenworth, Leavenworth, Kans	500
Fort McIntosh, Webb, Tex	710
Fort McKavett, Menard, Tex	483
Fort Quitman, El Paso, Tex	820
Fort Richardson, Jack, Tex	210
Fort Sill, Choctaw N., Ind. Ter	75
Fort Smith, Sebastian, Ark	194
Fort Stockton, Pecos, Tex	600
Fort Washita, Chickasaw N., Ind. Ter	92
Galveston, Galveston, Tex	507
Hillsborough, Hill, Tex	331
Houston, Harris, Tex	457
New Orleans, Orleans, La	809
San Antonio, Bexàr, Tex	599
Saint Louis, St. Louis, Mo	645
Seminole, Agency, Seminole N., Ind. Ter	81

FORT ARBUCKLE, IND. TER., TO—

	Miles.
Sherman, Grayson, Tex	129
Unita, Choctaw N., Ind. Ter	217
Waco, McLennan, Tex	366
Washington, D. C	1,551

FORT BARRANCAS, FLA., TO—

Fort Jefferson, Monroe, Fla	720
Key West, Monroe, Fla	656
New Orleans, Orleans, La	248
Pensacola, Escambia, Fla	8
Washington, D. C	1,024

FORT BASCOM, N. MEX., TO—

Fort Bayard, Grant, N. Mex	666
Fort Bliss, El Paso, Tex	614
Fort Craig, Socorro, N. Mex	440
Fort Cummings, Grant, N. Mex	619
Fort McRae, Socorro, N. Mex	472
Fort Reynolds, Pueblo, Colo	380
Fort Selden, Doña Aña, N. Mex	540
Fort Stanton, Socorro, N. Mex	204
Fort Sumner, San Miguel, N. Mex	80
Fort Union, Mora, N. Mex	145
Fort Wingate, Santa Aña, N. Mex., (new,)	459
Fort Wingate, Santa Aña, N. Mex., (old,)	407
Greenhorn, Huerfano, Colo	330
Pueblo, Pueblo, Colo	358
Santa Fé, Santa Fé, N. Mex	251
Trinidad, Las Animas, Colo	261
Washington, D. C	2,097

FORT BAYARD, N. MEX., TO—

Camp Bowie, Ariz	120
Fort Bliss, El Paso, Tex	158
Fort Craig, Socorro, N. Mex	226
Fort Cummings, Grant, N. Mex	45
Fort Garland, Costilla, Colo	578
Fort Leavenworth, Leavenworth, Kans	1,285
Fort McRae, Socorro, N. Mex	156
Fort Reynolds, Pueblo, Colo	745
Fort Selden, Doña Aña, N. Mex	126
Fort Stanton, Socorro, N. Mex	512
Fort Sumner, San Miguel, N. Mex	365
Fort Union, Union, N. Mex	521
Fort Wingate, Santa Aña, N. Mex., (new,)	481
Fort Wingate, Santa Aña, N. Mex., (old,)	429
Grenada, Colo	784

FORT BAYARD, N. MEX., TO—

	Miles.
Kit Carson, Greenwood, Colo	805
Las Cruces, Doña Aña, N. Mex	108
Mesilla, Doña Aña, N. Mex	105
Ralston, Grant, N. Mex	72
Rio Miembres, Grant, N. Mex	25
Santa Fé, Santa Fé, N. Mem	415
Sargent, Kans	797
Tucson, Pima, Ariz	232
Washington, D. C	2,473

FORT BENTON, MONT., TO—

Corinne, Box Elder, Utah	589
Cow Island, ———, Mont	147
Fort Buford, Howard, Dak	730
Fort Colville, Stevens, Wash	652
Fort Ellis, Gallatin, Mont	250
Fort Randall, Todd, Dak	1,446
Fort Rice, Morton, Dak	1,000
Fort Seward, Dak	1,075
Fort Shaw, Lewis and Clarke, Mont	64
Fort Stevenson, Stevens, Dak	880
Fort Sully, Sully, Dak	1,261
Fort Thompson, Buffalo, Dak	1,336
Fort Totten, Dak	1,155
Gallatin City, Gallatin, Mont	212
Grand River, Buffalo, Dak	1,170
Helena, Lewis and Clarke, Mont	144
New York, N. Y	2,657
Omaha, Douglas, Nebr	1,683
Ponka Agency, Todd, Dak	1,494
St. Paul, Ramsey, Minn	1,450
Sioux City, Woodbury, Iowa	1,577
Spokan Bridge, Stevens, Wash	632
Springville, Jefferson, Mont	177
Vermillion, Clay, Dak	1,546
Virginia City, Madison, Mont	267
Walla Walla, Walla Walla, Wash	787
Washington, D. C	2,588
Whetstone Agency, Gregory, Dak	1,432
Yankton, Yankton, Dak	1,519

FORT BLISS, TEX., TO—

Caddo, Ind. Ter	823
Dallas, Dallas, Tex	792
Denison, Tex	792
El Paso, El Paso, Tex	3
Fort Clark, Kinney, Tex	827
Fort Concho, Bexar, Tex	472
Fort Cummings, Grant, N. Mex	113
Fort Davis, Presidio, Tex	223
Fort Duncan, (Eagle Pass P. O.,) Maverick, Tex	872

FORT BLISS, TEX., TO—

	Miles.
Fort Gibson, Cherokee N., Ind. Ter	964
Fort Griffin, Shackelford, Tex	612
Fort Leavenworth, Kans	1,222
Fort McIntosh, Webb, Tex	866
Fort McKavett, Menard, Tex	527
Fort Quitman, El Paso, Tex	82
Fort Richardson, (Jacksborough P. O.,) Jack, Tex	692
Fort Selden, Doña Aña, N. Mex	74
Fort Sill, Choctaw N., Ind. Ter	988
Fort Stockton, Pecos, Tex	302
Galveston, Galveston, Tex	974
Harrisburgh, Harris, Tex	925
Houston, Harris, Tex	929
Indianola, Calhoun, Tex	859
Leasburgh, Doña Aña, N. Mex	73
Los Angeles, Los Angeles, Cal	965
Mesilla, Doña Aña, N. Mex	53
Ringgold Barracks, (Rio Grande City P. O.,) Starr, Tex	777
San Antonio, Bexar, Tex	701
San Bernardino, San Bernardino, Cal	940
San Francisco, San Francisco, Cal	1,429
Santa Fé, Santa Fé, N. Mex	363
Victoria, Victoria, Tex	813
Waco, McLennan, Tex	813
Washington, D. C	2,284

FORT BOISÉ, IDAHO, TO—

Fort Colville, Stevens, Wash	524
Fort Hall, Oneida, Idaho	457
Fort Klamath, Wasco, Oreg	535
Fort Lapwai, Nez Percés, Idaho	368
Fort Leavenworth, Leavenworth, Kans	1,554
Fort Stevens, Clatsop, Oreg	604
Kelton, Box Elder, Utah	250
Lewiston, Nez Percés, Idaho	356
Ogden City, Weber, Utah	342
Omaha, Douglas, Nebr	1,371
Portland, Multnomah, Oreg	500
Reno, Washoe, Nev	450
Sacramento, Sacramento, Cal	604
San Francisco, San Francisco, Cal	687
San Juan Island, Whatcom, Wash	752
Silver City, Owyhee, Idaho	70
Sitka, Alaska	1,575
Umatilla, Umatilla, Oreg	326
Vancouver, Clarke, Wash	432
Walla Walla, Walla Walla, Wash	271
Wallula, Walla Walla, Wash	301
Washington, D. C	2,640
Winnemucca, Humboldt, Nev	280

FORT BRADY, SAULT DE STE. MARIE, MICH., TO—

	Miles.
Detroit, Wayne, Mich	339
East Nubish Rapids, St. Mary's River, Mich	24
Fort Gratiot, (Port Huron Station,) St. Clair, Mich	276
Fort Howard, Brown, Wis	397
Fort Mackinac, Mackinac, Mich	92
Fort Wayne, Wayne, Mich	342
Houghton, Houghton, Mich	294
Marquette, Marquette, Mich	189
Menomonee, Menomonee, Mich	327
Milwaukee, Milwaukee, Wis	544
Oconto, Oconto, Wis	366
Ontonagon, Ontonagon, Mich	353
Peshtigo, Oconto, Wis	350
Port Huron, St. Clair, Mich	278
Superior City, Douglas, Wis	553
Sheboygan, Sheboygan, Wis	109
Washington, D. C	962
White Fish Point, Chippewa, Mich	42

FORT BRIDGER, WYO., TO—

Carter Station, Uintah, Wyo	11
Fort D. A. Russell, Laramie, Wyo	398
Fort Fetterman, Albany, Wyo	560
Fort Garland, Costilla, Colo	715
Fort Hall, Oneida, Idaho	302
Fort Kearney, Kearney, Nebr	726
Fort Laramie, Laramie, Wyo	476
Fort Leavenworth, Leavenworth, Kans	1,095
Fort Lyon, Bent, Colo	699
Fort McPherson, Lincoln, Nebr	641
Fort Reynolds, Pueblo, Colo	640
Fort Sedgewick, Weld, Colo	540
Fort Union, Mora, N. Mex	885
Fort Wallace, Wallace, Kans	714
Kit Carson, Greenwood, Colo	649
Laramie City, (Fort Sanders P. O.,) Albany, Wyo	452
Leavenworth City, Leavenworth, Kans	1,090
North Platte, Lincoln, Nebr	621
Ogden City, Weber, Utah	138
Omaha City, Douglas, Nebr	912
Pueblo, Pueblo, Colo	620
Salt Lake City, Salt Lake, Utah	175
San Francisco, San Francisco, Cal	964
Santa Fé, Santa Fé, N. Mex	991
Washington, D. C	2,176

FORT BUFORD, DAK., TO—

Fort Colville, Stevens, Wash	1,382
Fort Ellis, Gallatin, Mont	980
Fort Randall, Todd, Dak	716
Fort Rice, Morton, Dak	273
Fort Ripley, Morrison, Minn	661

FORT BUFORD, DAK., TO—

	Miles.
Fort Seward, Dak	345
Fort Shaw, Lewis and Clarke, Mont	794
Fort Stevenson, Stevens, Dak	150
Fort Sully, Sully, Dak	531
Fort Thompson, Buffalo, Dak	606
Fort Wadsworth, Deuel, Dak	517
Gallatin, Gallatin, Mont	942
Grand River, Buffalo, Dak	440
Helena, Lewis and Clarke, Mont	874
McCauleyville, Wilkin, Minn	440
New York, N Y	2,037
Omaha City, Douglas, Nebr	953
Ponka Agency, Todd, Dak	764
St. Paul, Ramsey, Minn	720
Sioux City, Woodbury, Iowa	847
Vermillion, Clay, Dak	814
Virginia City, Madison, Mont	997
Walla Walla, Walla Walla, Wash	1,517
Washington, D. C	1,908
Whetstone Agency, Gregory, Dak	704
Yankton, Yankton, Dak	789

FORT CLARK, TEX., TO—

Denison, Tex	606
Fort Concho, Bexar, Tex	355
Fort Davis, Presidio, Tex	827
Fort Duncan, Maverick, Tex	45
Fort Gibson, Cherokee N., Ind. Ter	776
Fort Leavenworth, Kans	1,036
Fort McIntosh, Webb, Tex	170
Fort McKavett, Menard, Tex	301
Fort Quitman, El Paso, Tex	745
Fort Richardson, Jack, Tex	575
Fort Sill, Choctaw N., Ind. Ter	800
Fort Stockton, Pecos, Tex	525
Galveston, Galveston, Tex	399
Houston, Harris, Tex	354
Indianola, Calhoun, Tex	234
New Orleans, Orleans, La	701
San Antonio, Bexar, Tex	126
Victoria, Victoria, Tex	238
Waco, McLennan, Tex	304
Washington, D. C	1,897
Weatherford, Parker, Tex	420

FORT COLUMBUS, N. Y., TO—

Albany, Albany, N. Y	145
Baltimore, Baltimore, Md	191
Boston, Suffolk, Mass	237
Burlington, Chittenden, Vt	292
Fort Foote, Prince George's, Md	239

FORT COLUMBUS, N. Y., TO—

	Miles.
Fort Hamilton, Kings, N. Y	9
Fort Independence, Suffolk, Mass	240
Fort Johnson, Brunswick, N. C	640
Fort Macon, Carteret, N. C	623
Fort McHenry, Baltimore, Md	194
Fort Preble, Cumberland, Me	346
Fort Sullivan, (Eastport,) Washington, Me	606
Fort Trumbull, New London, Conn	127
Fort Wadsworth, Richmond, N. Y	10
Fort Warren, Suffolk, Mass	244
Fort Washington, Prince George's, Md	244
Fort Wood, New York, N. Y	2
Newport, Newport, R. I	171
New York, New York, N. Y	1
Philadelphia, Philadelphia, Pa	90
Pikesville Arsenal, (Plattsburgh P. O.,) Baltimore, Md	199
Plattsburgh Barracks, Clinton, N. Y	314
Raleigh, Wake, N. C	545
Rutherfordton, Rutherford, N. C	723
Washington, D. C	229
Watertown Arsenal, Middlesex, Mass	245
Willett's Point, Queens, N. Y	21
Wilmington, Hanover, N. C	610

FORT COLVILLE, WASH., TO—

Deer Lodge City, Deer Lodge, Mont	453
Fort Ellis, Gallatin, Mont	614
Fort Klamath, Wasco, Oreg	923
Fort Lapwai, Nez Percés, Idaho	350
Fort Shaw, Lewis and Clarke, Mont	588
Fort Steilacoom, Pierce, Wash	692
Fort Stevens, Clatsop, Oreg	632
Helena, Lewis and Clarke, Mont	508
Jacksonville, Jackson, Oreg	823
Kelton, Box Elder, Utah	774
Lewiston, Nez Percés, Idaho	338
Marysville, Yuba, Cal	1,096
Missoula, Missoula, Mont	368
Monticello, Cowlitz, Wash	581
Ogden City, Weber, Utah	866
Olympia, Thurston, Wash	666
Portland, Multnomah, Oreg	528
Port Townsend, Jefferson, Wash	798
Red Bluff Tehama, Cal	1,010
Reno, Washoe, Nev	1,276
Sacramento, Sacramento, Cal	1,148
Sacramento Junction, Placer, Cal	1,130
San Francisco, San Francisco, Cal	1,207
San Juan Island, Whatcom, Wash	818
Sitka, Alaska	1,603
Spokan Bridge, Stevens, Wash	98
Umatilla, Umatilla, Oreg	308
Vancouver, Clarke, Wash	510
Walla Walla, Walla Walla, Wash	253
Washington, D. C	3,164
Winnemucca, Humboldt, Nev	804

FORT CONCHO, TEX., TO—

	Miles.
Camp Concho, ———, Tex	7
Dallas, Dallas, Tex	320
Denison, Tex	320
Fort Davis, Presidio, Tex	250
Fort Duncan, Maverick, Tex	400
Fort Gibson, Cherokee N., Ind. Ter	492
Fort Griffin, Shackelford, Tex	140
Fort Leavenworth, Kans	750
Fort McIntosh, Webb, Tex	394
Fort McKavett, Menard, Tex	55
Fort Quitman, El Paso, Tex	390
Fort Richardson, (Jacksborough P. O.,) Jack, Tex	220
Fort Sill, Choctaw N., Ind. Ter	516
Fort Stockton, Pecos, Tex	170
Galveston, Galveston, Tex	502
Hillsborough, Hill, Tex	341
Houston, Harris, Tex	453
Indianola, Calhoun, Tex	387
Mason, Mason, Tex	114
Menardville, Menard, Tex	76
New Orleans, Orleans, La	804
Ringgold Barracks, Starr, Tex	535
Rio Grande City, Starr, Tex	534
Saint Louis, St. Louis, Mo	906
San Antonio, Bexar, Tex	229
Victoria, Victoria, Tex	341
Waco, McLennan, Tex	341
Washington, D. C	1,812

FORT CRAIG, N. MEX., TO—

Fort Cummings, Grant, N. Mex	181
Fort Garland, Costilla, Colo	352
Fort Leavenworth, Leavenworth, Kans	1,059
Fort Lyon, Bent, Colo	449
Fort McRae, Socorro, N. Mex	32
Fort Reynolds, Pueblo, Colo	521
Fort Selden, Doña Aña, N. Mex	100
Fort Stanton, Socorro, N. Mex	286
Fort Sumner, San Miguel, N. Mex	291
Fort Union, Mora, N. Mex	295
Fort Wingate, Santa Aña, N. Mex., (new)	227
Grenada, Colo	558
Las Cruces, Doña Aña, N. Mex	118
Leasburgh, Doña Aña, N. Mex	101
Mesilla, Doña Aña, N. Mex	121
Santa Fé, Santo Fé, N. Mex	189
Washington, D. C	2,247

FORT CROSS, DAK.

See Fort Seward.

FORT CUMMINGS, N. MEX., TO

	Miles.
Camp Bowie, Ariz	165
Fort Dodge, Ford, Kan	1,093
Fort Garland, Costilla, Colo	533
Fort Leavenworth, Kans	1,240
Fort Lyon, Bent, Colo	610
Fort McRae, Socorro, N. Mex	111
Fort Reynolds, Pueblo, Colo	702
Fort Selden, Doña Aña, N. Mex	81
Fort Stanton, Socorro, N. Mex	467
Fort Union, Moro, N. Mex	476
Fort Wingate, Santa Aña, N. Mex	365
Fort Yuma, San Diego, Cal	565
Grenada, Colo	739
Maricopa Wells, Pima, Ariz	355
Mesilla, Doña Aña, N. Mex	60
Prescott, Yarapai, Ariz	588
Ralston, Grant, M. Mex	117
Rio Miembres, Grant, N. Mex	20
San Francisco, San Francisco, Cal	1,316
Santa Fé, Santa Fé, N. Mex	370
Tucson, Pima, Ariz	277
Washington, D. C	2,428

FORT D. A. RUSSELL, WYO., TO—

Fort Fetterman, Albany, Dak	172
Fort Fred Steele, Carbon, Wyo	183
Fort Garland, Costilla, Colo	321
Fort Hall, Oneida, Idaho	679
Fort Kearney, Kearney, Nebr	332
Fort Laramie, Laramie, Wyo	82
Fort Leavenworth, Leavenworth, Kans	699
Fort Lyon, Bent, Colo	305
Fort McPherson, Lincoln, Nebr	247
Fort Reynolds, Pueblo, Colo	246
Fort Sedgwick, Wild, Colo	144
Fort Wallace, Wallace, Kans	320
Julesburgh, Wild, Colo	141
Kit Carson, Greenwood, Colo	255
Laramie City, Albany, Wyo	58
Lawrence, Douglas, Kans	704
Leavenworth City, Leavenworth, Kans	696
Ogden City, Weber, Utah	515
Omaha City, Douglas, Nebr	518
Pueblo, Pueblo, Colo	226
Salt Lake City, Salt Lake, Utah	552
San Francisco, San Francisco, Cal	1,341
Santa Fé, Santa Fé. N. Mex	597
Washington, D. D	1,784

FORT DAVIS, TEX., TO—

	Miles.
Dallas, Dallas, Tex	570
Denison, Tex	570
Fort Duncan, (Eagle Pass P. O.,) Maverick, Tex	650
Fort Gibson, Cherokee, N., Ind. Ter	742
Fort Griffin, Shackelford, Tex	390
Fort Leavenworth, Kans	1,000
Fort McIntosh, Webb, Tex	644
Fort McKavett, Menard, Tex	303
Fort Quitman, El Paso, Tex	140
Fort Richardson, (Jacksborough P. O.,) Jack, Tex	470
Fort Selden, Doña Aña, N. Mex	296
Fort Sill, Choctaw N., Ind. Ter	766
Fort Stockton, Pocos, Tex	80
Galveston, Galveston, Tex	753
Houston, Harris, Tex	708
Las Cruces, Doña Aña, N. Mex	278
Leasburgh, Doña Aña, N. Mex	295
Los Angeles, Los Angeles, Cal	1,224
Mesilla, Doña Aña, N. Mex	275
New Orleans, Orleans, La	1,055
Presidio del Norte, Mexico	110
Ringgold Barracks, Starr, Tex	785
San Antonio, Bexar, Tex	479
San Bernardino, San Bernardino, Cal	1,162
San Diego, San Diego, Cal	1,134
San Francisco, San Francisco, Cal	1,651
Santa Fé, Santa Fé, N. Mex	585
Victoria, Victoria, Tex	591
Waco, McLennan, Tex	591
Washington, D. C	2,110

FORT DODGE, KANS., TO—

Dodge City, Kans	4
Emporia, Lyon, Kans	242
Fort Gibson, Cherokee N., Ind. Ter	455
Fort Harker, Ellsworth, Kans	138
Fort Hays, Ellis, Kans	80
Fort Larned, Pawnee, Kans	65
Fort Leavenworth, Leavenworth, Kans	370
Fort Union, Mora, N. Mex	597
Fort Wingate, Santa Aña, N. Mex	931
Kansas City, Jackson, Mo	369
Kit Carson, Greenwood, Colo	279
Saint Louis, Saint Louis, Mo	652
Santa Fé, Santa Fé, N. Mex	721
Trinidad, Las Animas, Colo	447
Washington, D. C	1,559

FORT DUNCAN, TEX., TO—

Fort Gibson, Cherokee N., Ind. Ter	826
Fort Griffin, Shackelford, Tex	540
Fort McIntosh, Webb, Tex	125

FORT DUNCAN, TEX., TO—

	Miles.
Fort McKavett, Menard, Tex	353
Fort Quitman, El Paso, Tex	797
Fort Richardson, (Jacksborough P. O.,) Jack, Tex	627
Fort Sill, Choctaw N., Ind. Ter	850
Fort Stockton, Pecos, Tex	577
Galveston, Galveston, Tex	452
Houston, Harris, Tex	407
Indianola, Calhoun, Tex	336
New Orleans, Orleans, La	759
Ringgold Barracks, (Rio Grande City,) Starr, Tex	265
San Antonia, Bexar, Tex	171
Uvalde, Uvalde, Tex	83
Victoria, Victoria, Tex	290
Waco, McLennan, Tex	354
Washington, D. C	1,950

FORT ELLIS, MONT., TO—

Bozeman, Gallatin, Mont	3
Fort Laramie Laramie, Wyo	1,073
Fort Randall, Todd, Dak	1,575
Fort Rice, Morton, Dak	1,253
Fort Ripley, Morrison, Minn	1,641
Fort Seward, Dak	1,325
Fort Shaw, Lewis and Clarke, Mont	186
Fort Stevenson, Stevens, Dak	1,130
Fort Sully, Sully, Dak	1,511
Fort Thompson, Buffalo, Dak	1,586
Gallatin, Clay, Mont	38
Grand River, Buffalo, Dak	1,420
Helena, Lewis and Clarke, Mont	106
Mouth of Mussel Shell River, Dawson, Mont	249
Ogden City, Weber, Utah	480
Omaha City, Douglas, Nebr	1,509
Ponka Agency, Todd, Dak	1,744
Saint Paul, Ramsey, Minn	1,660
Salt Lake City, Salt Lake, Utah	517
Sioux City, Woodbury, Iowa	1,609
Vermillion, Dakota, Minn	1,695
Virginia City, Madison, Mont	78
Washington, D. C	2,775
Whetstone Agency, Gregory, Dak	1,682
Yankton, Yankton, Dak	1,676

FORT FOOTE, MD., TO—

Alexandria, Alexandria, Va	2
Baltimore, Baltimore, Md	48
Boston, Suffolk, Mass	474
Fort Hamilton, Kings, N. Y	246
Fort Independence, Suffolk, Mass	477
Fort Johnson, Brunswick, N. C	401
Fort Macon, Carteret, N. C	384

FORT FOOTE, MD., TO—

	Miles.
Fort McHenry, Baltimore, Md	51
Fort Preble, Cumberland, Md	673
Fort Sullivan, Washington, Me	843
Fort Trumbull, New London, Conn	364
Fort Wadsworth, Richmond, N. Y	247
Fort Warren, Suffolk, Mass	481
Fort Washington, Prince George's, Md	7
Fort Wood, Wood, N. Y	240
Newport, Newport, R. I	408
New York, New York, N. Y	238
Philadelphia, Philadelphia, Pa	149
Pikesville Arsenal, Baltimore, Md	56
Plattsburgh Barracks, (Plattsburgh P. O.,) Clinton, N. Y	551
Raleigh, Wake, N. C	306
Richmond, Henrico, Va	121
Rutherfordton, Rutherford, N. C	484
Washington, D. C	10
Watertown Arsenal, Middlesex, Mass	482
Willett's Point, Queens, N. Y	258
Wilmington, New Hanover, N. C	371

FORT FRED STEELE, WYO., TO

Fort Garland, Costilla, Colo	500
Fort Hall, Oneida, Idaho	499
Fort Kearney, Kearney, Nebr	511
Fort Laramie, Laramie, Wyo	261
Fort Leavenworth, Leavenworth, Kans	878
Fort Lyon, Bent, Colo	484
Fort McPherson, Lincoln, Nebr	426
Fort Reynolds, Pueblo, Colo	425
Fort Sedgwick, Weld, Colo	323
Fort Steele Station, Carbon, Wyo	2
Fort Wallace, Wallace, Kans	499
Julesburgh, Weld, Colo	320
Kit Carson, Greenwood, Colo	434
Laramie City, Albany, Wyo	125
Lawrence, Douglas, Kans	883
Leavenworth City, Leavenworth, Kans	875
Ogden City, Weber, Utah	335
Omaha City, Douglas, Nebr	697
Pueblo, Pueblo, Colo	405
Salt Lake City, Salt Lake, Utah	372
San Francisco, San Francisco, Cal	1,161
Santa Fé, Santa Fé, N. Mex	776
Trinidad, Las Animas, Colo	502
Washington, D. C	1,963

FORT GARLAND, COLO., TO—

Badito, Huerfano, Colo	32
Bent's Fort, Bent, Colo	175
Fernandez de Taos, Taos, N. Mex	85
Fort Harker, Ellsworth, Kans	522
Fort Hays, Ellis, Kans	442
Fort Kearney, Kearney, Nebr	649

FORT GARLAND, COLO., TO—

	Miles.
Fort Laramie, Laramie, Wyo	399
Fort Larned, Pawnee, Kans	483
Fort Leavenworth, Leavenworth, Kans	728
Fort Lyon, Bent, Colo	193
Fort McPherson, Lincoln, Nebr	564
Fort McRae, Socorro, N. Mex	384
Fort Reynolds, Pueblo, Colo	117
Fort Sedgwick, Weld, Colo	461
Fort Selden, Doña Aña, N. Mex	452
Fort Stanton, Socorro, N. Mex	402
Fort Union, Mora, N. Mex	304
Fort Wallace, Wallace, Kans	308
Fort Wingate, Santa Aña, N. Mex	371
Greenhorn, Huerfano, Colo	67
Kansas City, Jackson, Mo	731
Kit Carson, Greenwood, Colo	243
Leavenworth City, Leavenworth, Kans	725
Mesilla, Doña Aña, N. Mex	473
Pueblo, Pueblo, Colo	95
Santa Fé, Santa Fé, N. Mex	163
South Side, Bent, Colo	137
Trinidad, Las Animas, Colo	136
Washington, D. C	1,920

FORT GIBSON, CHEROKEE N., IND. TER., TO—

Dallas, Dallas, Tex	244
Denison, Tex	173
Emporia, Lyon, Kans	213
Fort Griffin, Shackelford, Tex	353
Fort Harker, Ellsworth, Kans	354
Fort Larned, Pawnee, Kans	398
Fort Leavenworth, Leavenworth, Kans	275
Fort McIntosh, Webb, Tex	875
Fort McKavett, Menard, Tex	658
Fort Richardson, (Jacksborough P. O.,) Jack, Tex	273
Fort Scott, Bourbon, Kans	167
Fort Sill, Choctaw N., Ind. Ter	307
Fort Smith, Sebastian, Ark	75
Fort Stockton, Pecos, Tex	663
Galveston, Galveston, Tex	559
Gibson Station, Cherokee N., Ind. Ter	8
Grenada, Colo	583
Houston, Harris, Tex	550
Humboldt, Allen, Kans	153
Indianola, Calhoun, Tex	620
Junction City, Davis, Kans	274
Kansas City, Jackson, Mo	265
Laredo, (Fort McIntosh,) Webb, Tex	815
Larned, Pawnee, Kans	394
Lawrence, Douglas, Kans	239
Leavenworth City, Leavenworth, Kans	272
Parsons, Kans	118
San Antonio, Bexar, Tex	650
Saint Louis, Saint Louis, Mo	432
Vinita, Cherokee N., Ind. Ter	67
Washington, D. C	1,338

FORT GRATIOT, MICH., TO—

	Miles.
Fort Mackinac, Mackinac, Mich	240
Fort Niagara, Niagara, N. Y	195
Fort Wayne, Wayne, Mich	66
Marquette, Marquette, Mich	465
Milwaukee, Milwaukee, Wis	288
Ontonagon, Ontonagon, Mich	629
Port Huron, St. Clair, Mich	1
Superior, Douglas, Wis	766
Washington, D. C	686
White Fish Point, Chippewa, Mich	318

FORT GRIFFIN, TEX., TO—

Fort Leavenworth, Leavenworth, Kans	604
Fort McIntosh, Webb, Tex	534
Fort McKavett, Menard, Tex	195
Fort Quitman, El Paso, Tex	530
Fort Richardson, (Jacksborough P. O.,) Jack, Tex	80
Fort Sill, Choctaw N., Ind. Ter	376
Fort Smith, Sebastian, Ark	484
Fort Stockton, Pecos, Tex	310
Galveston, Galveston, Tex	479
Hillsborough, Hill, Tex	201
Houston, Harris, Tex	428
Indianola, Calhoun, Tex	516
Jacksborough, (Fort Richardson,) Jack, Tex	81
Laredo, (Fort McIntosh,) Webb, Tex	534
New Orleans, Orleans, La	781
Orange, Orange, Tex	538
Ringgold Barracks, Starr, Tex	717
Rio Grande City, Starr, Tex	716
San Antonio, Bexar, Tex	369
Victoria, Victoria, Tex	470
Waco, McLennan, Tex	237
Washington, D. C	1,778

FORT HALL, IDAHO, TO—

Fort Harker, Ellsworth, Kans	1,199
Fort Hays, Ellis, Kans	1,129
Fort Klamath, Wasco, Oreg	1,098
Fort Leavenworth, Leavenworth, Kans	1,615
Fort Lyon, Bent, Colo	980
Fort McPherson, Lincoln, Nebr	922
Fort Reynolds, Pueblo, Colo	921
Fort Sedgwick, Weld, Colo	819
Fort Steilacoom, Pierce, Wash	1,121
Fort Stevens, Clatsop, Oreg	1,061
Jacksonville, Jackson, Oreg	1,148
Kelton, Box Elder, Utah	207
Monticello, Cowlitz, Wash	1,010
Ogden City, Weber, Utah	164
Olympia, Thurston, Wash	1,095

FORT HALL, IDAHO, TO—

	Miles.
Omaha City, Douglas, Nebr	1,193
Portland, Multnomah, Oreg	957
Port Townsend, Jefferson, Wash	1,227
Red Bluff, Tehama, Cal	961
Reno, Washoe, Nev	705
Ross Fork, Oneida, Idaho	15
Sacramento, Sacramento, Cal	859
Sacramento Junction, Placer, Cal	841
Salt Lake City, Salt Lake, Utah	201
San Francisco, San Francisco, Cal	942
San Juan Island, Whatcom, Wash	1,209
Sitka, Alaska	2,032
Umatilla, Umatilla, Oreg	783
Vancouver, Clarke, Wash	889
Walla Walla, Walla Walla, Wash	728
Wallula, Walla Walla, Wash	758
Washington, D. C	2,462
Winnemucca, Humboldt, Nev	535

FORTS HAMILTON AND LA FAYETTE, N. Y., TO—

Albany, Albany, N. Y	142
Baltimore, Baltimore, Md	198
Boston, Suffolk, Mass	244
Fort Independence, Suffolk, Mass	246
Fort Johnson, Brunswick, N. C	645
Fort Macon, Carteret, N. C	628
Fort McHenry, Baltimore, Md	201
Fort Monroe, Elizabeth City, Va	383
Fort Preble, Cumberland, Me	353
Fort Sullivan, (Eastport,) Washington, Me	613
Fort Trumbull, New London, Conn	134
Fort Wadsworth, Richmond, N. Y	1
Fort Warren, Suffolk, Mass	251
Fort Washington, Prince George's, Md	251
Fort Wood, New York, N. Y	10
Morehead City, Carteret, N. C	626
Newport, Newport, R. I	178
New York, New York, N. Y	8
Philadelphia, Philadelphia, Pa	97
Pikesville Arsenal, Baltimore, Md	206
Plattsburgh Barracks, Clinton, N. Y	321
Raleigh, Wake, N. C	552
Rutherfordton, Rutherford, N. C	730
Washington, D. C	236
Watertown Arsenal, Middlesex, Mass	252
Willett's Point, Queens, N. Y	28
Wilmington, New Hanover, N. C	615

FORT HARKER, KANS., TO—

Fort Hays, Ellis, Kans	70
Fort Laramie, Laramie, Wyo	602
Fort Larned, Pawnee, Kans	116
Fort Leavenworth, Leavenworth, Kans	216

FORT HARKER, KANS., TO—

	Miles.
Fort Lyon, Bent, Colo	319
Fort Reynolds, Pueblo, Colo	395
Fort Riley, Davis, Kans	83
Fort Sedgwick, Weld, Colo	664
Fort Sill, Choctaw N., Ind. Ter	654
Fort Union, Mora, N. Mex	553
Fort Wallace, Wallace, Kans	204
Kansas City, Jackson, Mo	220
Kit Carson, Greenwood, Colo	269
Laramie City, Albany, Wyo	578
Leavenworth City, Leavenworth, Kans	213
Ogden City, Weber, Utah	1,035
Saint Louis, Saint Louis, Mo	502
Salt Lake City, Salt Lake, Utah	1,072
Santa Fé, Santa Fé, N. Mex	711
Washington, D. C	1,409

FORT HAYS, KANS., TO—

Fort Laramie, Laramie, Wyo	532
Fort Larned, Pawnee, Kans	46
Fort Leavenworth, Leavenworth, Kans	286
Fort Lyon, Bent, Colo	249
Fort Reynolds, Pueblo, Colo	325
Fort Riley, Davis, Kans	153
Fort Union, Mora, N. Mex	465
Fort Wallace, Wallace, Kans	134
Kansas City, Jackson, Mo	290
Kit Carson, Greenwood, Colo	199
Laramie City, Albany, Wyo	508
Leavenworth City, Leavenworth, Kans	283
Ogden City, Weber, Utah	965
Saint Louis, Saint Louis, Mo	572
Salt Lake City, Salt Lake, Utah	1,002
Santa Fé, Santa Fé, N. Mex	623
Washington, D. C	1,479

FORT INDEPENDENCE, MASS., TO—

Baltimore, Baltimore, Md	428
Boston, Suffolk, Mass	2½
City Point, Suffolk, Mass	½
Fort Johnson, Brunswick, N. C	877
Fort Macon, Carteret, N. C	858
Fort McHenry, Baltimore, Md	431
Fort Monroe, Elizabeth City, Va	613
Port Preble, Cumberland, Me	111
Fort Sullivan, Washington, Me	371
Fort Trumbull, New London, Conn	113
Fort Wadsworth, Richmond, N. Y	247
Fort Warren, Suffolk, Mass	4
Fort Washington, Prince George's, Md	481
Fort Winthrop, Suffolk, Mass	1

FORT INDEPENDENCE, MASS., TO—

	Miles.
Fort Wood, New York, N. Y	240
Newport, Newport, R. I	71
New York, New York, N. Y	238
Philadelphia, Philadelphia, Pa	328
Pikesville Arsenal, Baltimore, Md	436
Plattsburgh Barracks, Clinton, N. Y	258
Raleigh, Wake, N. C	782
Rutherfordton, Rutherford, N. C	960
Washington, D. C	466
Watertown Arsenal, Middlesex, Mass	10½
Willett's Point, Queens, N. Y	258
Wilmington, New Hanover, N. C	847

FORTS JACKSON AND SAINT PHILIP, LA., TO—

Baton Rouge, East Baton Rouge, La	204
Jackson Barracks, Orleans, La	77
Mount Vernon, Mobile, Ala	259
New Orleans, Orleans, La	74
Washington, D. C	1,261

FORT JEFFERSON, (TORTUGAS,) FLA., TO—

Baltimore, Baltimore, Md	1,164
Cedar Keys, Levy, Fla	394
Key West, Monroe, Fla	64
New Orleans, Orleans, La	664
Washington, D. C	1,202

FORT JOHNSON, N. C., TO—

Charlotte, Mecklenburgh, N. C	226
Cherryville, Gaston, N. C	269
Fort Macon, Carteret, N. C	211
Fort McHenry, Baltimore, Md	452
Fort Monroe, Elizabeth City, Va	286
Fort Preble, Cumberland, Me	984
Fort Sullivan, Washington, Me	1,244
Fort Trumbull, New London, Conn	765
Fort Wadsworth, Richmond, N. Y	648
Fort Washington, Prince George's, Md	396
Fort Warren, Suffolk, Mass	882
Fort Wood, New York, N. Y	641
Lylesville, N. C	162
Morehead City, Carteret, N. C	209
New Haven, New Haven, Conn	715
New London, New London, Conn	765
New York, New York, N. Y	639
Philadelphia, Philadelphia, Pa	550
Pikesville Arsenal, Baltimore, Md	457

FORT JOHNSON, N. C., TO—

	Miles.
Plattsburgh Barracks, Clinton, N. Y	952
Raleigh, Wake, N. C	162
Rockingham, Richmond, N. C	147
Rutherfordton, Rutherford, N. C	306
Washington, D. C	411
Watertown Arsenal, Middlesex, Mass	883
Willett's Point, Queens, N. Y	659
Wilmington, New Hanover, N. C	30

FORT KEARNEY, NEBR., TO—

Fort Laramie, Laramie, Wyo	410
Fort Leavenworth, Leavenworth, Kans	379
Fort McPherson, Lincoln, Nebr	97
Fort Reynolds, Pueblo, Colo	574
Fort Sedgwick, Weld, Colo	194
Kearney City, Kearney, Nev	5
Laramie City, Albany, Wyo	386
North Platte, Lincoln, Nebr	105
Ogden City, Weber, Utah	843
Omaha City, Douglas, Nebr	196
Salt Lake City, Salt Lake, Utah	880
Sidney, Cheyenne, Nebr	228
Washington, D. C	1,462

FORT KLAMATH, OREG., TO—

Fort Lapwai, Nez Percés, Idaho	765
Fort Leavenworth, Leavenworth, Kans	2,194
Fort Steilacoom, Pierce, Wash	559
Fort Stevens, Clatsop, Oreg	497
Jacksonville, Jackson, Oreg	100
Lewiston, Nez Percés, Idaho	755
Marysville, Yuba, Cal	373
Monticello, Cowlitz, Wash	448
Olympia, Thurston, Wash	533
Omaha City, Douglas, Nebr	2,011
Portland, Multnomah, Oreg	395
Port Townsend, Jefferson, Wash	665
Red Bluff, Tehama, Cal	287
Reno, Washoe, Nev	393
Sacramento, Sacramento, Cal	425
Sacramento Junction, Placer, Cal	407
San Francisco, San Francisco, Cal	484
San Juan Island, Whatcom, Wash	695
Sitka, Alaska	1,470
Umatilla, Umatilla, Oreg	615
Vancouver, Clarke, Wash	413
Victoria, Brit. Col	715
Walla Walla, Walla Walla, Wash	670
Wallula, Walla Walla, Wash	640
Washington, D. C	3,579

FORT LAPWAI, IDAHO, TO—

	Miles.
Fort Leavenworth, Leavenworth, Kans	1,922
Fort Steilacoom, Pierce, Wash	539
Fort Stevens, Clatsop, Oreg	476
Jacksonville, Jackson, Oreg	667
Kelton, Box Elder, Utah	618
Lewiston, Nez Percés, Idaho	12
Marysville, Yuba, Cal	940
Monticello, Cowlitz, Wash	428
Ogden City, Weber, Utah	710
Olympia, Thurston, Wash	513
Omaha City, Douglas, Nebr	1,739
Portland, Multnomah, Oreg	372
Port Townsend, Jefferson, Wash	645
Red Bluff, Temaha, Cal	854
Reno, Washoe, Nev	818
Sacramento, Sacramento, Cal	992
Sacramento Junction, Placer, Cal	974
San Francisco, San Francisco, Cal	1,051
San Juan Island, Whatcom, Wash	675
Sitka, Alaska	1,447
Umatilla, Umatilla, Oreg	152
Vancouver, Clarke, Wash	355
Victoria, Brit. Col	655
Walla Walla, Walla Walla, Wash	97
Wallula, Walla Walla, Wash	127
Washington, D. C	3,008
Winnemucca, Humboldt, Nev	648

FORT LARAMIE, WYO., TO—

Fort Leavenworth, Leavenworth, Kans	777
Fort Lyon, Bent, Colo	383
Fort McPherson, Lincoln, Nebr	325
Fort Reynolds, Pueblo, Colo	324
Fort Sanders, (Laramie City,) Albany, Wyo	134
Fort Sedgwick, Weld, Colo	222
Wallace, Wallace, Kans	398
North Platte, Lincoln, Nebr	305
Ogden City, Weber, Utah	593
Omaha City, Douglas, Nebr	596
Pueblo, Pueblo, Colo	304
Salt Lake City, Salt Lake, Utah	630
Santa Fé, Santa Fé, N. Mex	675
Sidney, Cheyenne, Nebr	182
Trinidad, Las Animas, Colo	401
Washington, D. C	1,862

FORT LARNED, KANS., TO—

Fort Leavenworth, Leavenworth, Kans	314
Fort Lyon, Bent, Colo	295
Fort Reynolds, Pueblo, Colo	371
Fort Union, Mora, N. Mex	563

FORT LARNED, KANS., TO—

	Miles.
Fort Wallace, Wallace, Kans	180
Fort Zarah, Barton, Kans	32
Grenada, Colo	180
Kansas City, Jackson, Mo	314
Kit Carson, Greenwood, Colo	245
Laramie City, Albany, Wyo	554
Leavenworth City, Leavenworth, Kans	317
North Topeka, Shawnee, Kans	247
Ogden City, Weber, Utah	1,011
Saint Louis, Saint Louis, Mo	598
Salt Lake City, Salt Lake, Utah	1,048
Santà Fé, Santa Fé, N. Mex	562
Washington, D. C	1,525

FORT LEAVENWORTH, KANS., TO—

Albany, Albany, N. Y	1,317
Albuquerque, Bernalillo, N. Mex	941
Alton, Madison, Ill	337
Annapolis, Anne Arundel, Md	1,260
Atchison, Atchison, Kans	18
Atlanta, Fulton, Ga	923
Augusta, Richmond, Ga	1,094
Augusta, Kennebec, Me	1,692
Austin, Travis, Tex	831
Baltimore, Baltimore, Md	1,243
Bangor, Penobscot, Me	1,765
Bath, Sagadahoc, Me	1,663
Baton Rouge, East Baton Rouge, La	1,158
Beaufort, Carteret, N. C	1,645
Benicia, Solano, Cal	2,088
Boisé City, Ada, Idaho	1,554
Boston, Suffolk, Mass	1,519
Brattleborough, Windham, Vt	1,479
Buffalo, Erie, N. Y	1,018
Burlington, Des Moines, Iowa	342
Burlington, Crittenden, Vt	1,474
Cairo, Alexandria, Ill	459
Camp Baker, Meagher, Mont	1,725
Camp Douglas, Salt Lake, Utah	1,251
Camp Halleck, Elko, Nev	1,517
Camp Harney, Oreg	1,896
Camp McDermitt, Humboldt, Nev	1,711
Camp Stambaugh, Sweetwater, Wyo	1,493
Camp Supply, ——, Ind. Ter	461
Camp Three Forks, Owyhee, Idaho	1,659
Camp Warner, Grant, Oreg	2,105
Canyon City, Grant, Oreg	1,554
Carlisle, Cumberland, Penn	1,196
Charleston, Charleston, S. C	1,230
Chattanooga, Hamilton, Tenn	784
Cheyenne, Laramie, Wyo	699
Chicago, Cook, Ill	516
Cincinnati, Hamilton, Ohio	653
Cleveland, Cuyahoga, Ohio	833
Columbia, Richland, S. C	1,179

FORT LEAVENWORTH, KANS., TO—

	Miles.
Columbus, Franklin, Ohio	737
Concord, Merrimack, N. H	1,557
Corinne, Box Elder, Utah	1,220
Davenport, Scott, Iowa	350
Dayton, Montgomery, Ohio	659
Denver, Arapahoe, Colo	636
Des Moines, Polk, Iowa	312
Detroit, Wayne, Mich	795
Dover, Kent, Del	1,361
Elko, Elko, Nev	1,487
Elmiria, Chemung, N. Y	1,164
Erie, Erie, Pa	930
Evansville, Vanderburgh, Ind	473
Fernandina, Nassau, Fla	1,405
Fort Bayard, Grant, N. Mex	1,285
Fort Benton, Chouteau, Mont	1,809
Fort Bliss, El Paso, Tex	1,222
Fort Boisé, Ada, Idaho	1,554
Fort Bridger, Unitah, Wyo	1,095
Fort Clark, Kinney, Tex	1,036
Fort Colville, Stevens, Wash	2,078
Fort Concho, Bexar, Tex	750
Fort Craig, Socorro, N. Mex	1,059
Fort Cummings, Grant, N. Mex	1,240
Fort D. A. Russell, Laramie, Wyo	699
Fort Davis, Presidio, Tex	1,000
Fort Dodge, Ford, Kans	370
Fort Ellis, Gallatin, Mont	1,676
Fort Fred Steele, Carbon, Wyo	878
Fort Garland, Costilla, Colo	728
Fort Gibson, Cherokee N., Ind. Ter	275
Fort Griffin, Shackelford, Tex	604
Fort Hall, Oneida, Idaho	1,615
Fort Harker, Ellsworth, Kans	216
Fort Hays, Ellis, Kans	286
Fort Howard, Brown, Wis	758
Fort Kearney, Kearney, Nebr	379
Fort Klamath, Wasco, Oreg	2,194
Fort Lapwai, Nez Percés, Idaho	1,922
Fort Laramie, Laramie, Wyo	777
Fort Larned, Pawnee, Kans	314
Fort Lyon, Bent, Colo	535
Fort McPherson, Lincoln, Nebr	460
Fort McRae, Socorro, N. Mex	1,091
Fort Monroe, Elizabeth City, Va	1,428
Fort Quitman, El Paso, Tex	1,140
Fort Reynolds, Pueblo, Colo	611
Fort Richardson, Jack, Tex	530
Fort Sanders, Albany, Wyo	747
Fort Sedgwick, Weld, Colo	557
Fort Selden, Doña Aña, N. Mex	1,159
Fort Shaw, Lewis and Clarke, Mont	1,745
Fort Sill, Choctaw N., Ind. Ter	575
Fort Smith, Sebastian, Ark	350
Fort Snelling, Hennepin, Minn	536
Fort Steilacoom, Pierce, Wash	2,218
Fort Stevens, Clatsop, Oreg	2,158
Fort Stockton, Pecos, Tex	920
Fort Union, Mora, N. Mex	762
Fort Wallace, Wallace, Kans	420

FORT LEAVENWORTH, KANS., TO—

	Miles.
Fort Wayne, Allen, Ind	636
Fort Wingate, Santa Aña, N. Mex., (new)	1,078
Fort Wingate, Santa Aña, N. Mex., (old)	1,026
Frankfort, Franklin, Ky	654
Galveston, Galveston, Tex	816
Grand Haven, Ottawa, Mich	701
Houston, Harris, Tex	766
Indianapolis, Marion, Ind	550
Jackson, Hinds, Miss	863
Jacksonville, Duval, Fla	1,378
Jeffersonville, Clarke, Ind	590
Joliet, Will, Ill	493
Kansas City, Jackson, Mo	29
Kelton, Box Elder, Utah	1,304
Keokuk, Lee, Iowa,	294
Kit Carson, Greenwood, Colo	485
Knoxville, Knox, Tenn	896
La Crosse, La Crosse, Wis	644
La Fayette, Tippecanoe, Ind	527
Lansing, Ingham, Mich	722
Lawrence, Douglas, Kans	36
Leavenworth City, Leavenworth, Kans	3
Logansport Cass, Ind	564
Louisville, Jefferson, Ky	590
Lynchburgh, Campbell, Va	1,231
Macon, Bibb, Ga	1,026
Madison, Dane, Wis	648
Marietta, Washington, Ohio	852
Memphis, Shelby, Tenn	658
Michigan City, La Porte, Ind	572
Milledgeville, Baldwin, Ga	1,065
Mobile, Mobile, Ala	979
Montgomery, Montgomery, Ala	995
Monticello, Cowlitz, Wash	2,107
Nashville, Davidson, Tenn	690
New Haven, New Haven, Conn	1,438
New London, New London, Conn	1,488
New Orleans, Orleans, La	1,010
New York, New York, N. Y	1,362
Norfolk, Norfolk, Va	1,439
Ogden City, Weber, Utah	1,212
Ogdensburgh, St. Lawrence, N. Y	1,308
Olympia, Thurston, Wash	2,192
Omaha, Douglas, Nebr	183
Oswego, Oswego, N. Y	1,204
Parkersburgh, Wood, Va	860
Pensacola, Escambia, Fla	1,090
Philadelphia, Philadelphia, Pa	1,286
Pittsburgh, Allegheny, Pa	929
Port Huron, St. Clair, Mich	857
Portland, Cumberland, Me	1,627
Portland, Multnomah, Oreg	2,054
Port Townsend, Jefferson, Wash	2,324
Prairie du Chien, Crawford, Wis	580
Providence, Providence, R. I	1,553
Quincy, Adams, Ill	254
Raleigh, Wake, N. C	1,465
Reading, Berks, Pa	1,232
Red Bluff, Tehama, Cal	2,059
Reno, Washoe, Nev	1,801

FORT LEAVENWORTH, KANS., TO—

	Miles.
Richmond, Wayne, Ind	570
Richmond, Henrico, Va	1,382
Rock Island, Rock Island, Ill	350
Rouse's Point, Clinton, N. Y	1,426
Sacramento, Sacramento, Cal	1,952
St. Joseph, Buchanan, Mo	50
St. Louis, St. Louis, Mo	311
St. Paul, Ramsey, Minn	528
San Antonio, Bexar, Tex	908
Sandusky, Erie, Ohio	767
San Francisco, San Francisco, Cal	2,039
San Juan Island, Whatcom, Wash	2,354
Santa Fé, Santa Fé, N. Mex	870
Savannah, Chatham, Ga	1,226
Sioux City, Woodbury, Iowa	258
Springfield, Sangamon, Ill	356
Springfield, Hampden, Mass	1,420
Stevenson, Jackson, Ala	745
Syracuse, Onondaga, N. Y	1,168
Tallahassee, Leon, Fla	1,377
Terre Haute, Vigo, Ind	477
The Dalles, Wasco, Oreg	1,934
Toledo, Lucas, Ohio	718
Topeka, Shawnee, Kans	63
Vicksburgh, Warren, Miss	898
Victoria, Wirt, Colo	2,334
Washington, D. C	1,219
Weldon, Halifax, N. C	1,422
West Point, Orange, N. Y	1,418
Wheeling, Ohio, W. Va	880
Wilmington, New Castle, Del	1,314
Wilmington, New Hanover, N. C	1,369
Winona, Winona, Minn	676

FORT LYON, COLO., TO—

Bent's Fort, Bent, Colo	18
Fort Lyon, (old,) Bent, Colo	21
Fort McPherson, Lincoln, Nebr	548
Fort McRae, Socorro, N. Mex	516
Fort Reynolds, Pueblo, Colo	78
Fort Sanders, Albany, Wyo	357
Fort Sedgwick, Weld, Colo	445
Fort Selden, Doña Aña, N. Mex	681
Fort Sumner, San Miguel, N. Mex	382
Fort Union, Mora, N. Mex	234
Fort Wallace, Wallace, Kans	115
Fort Wingate, Santa Aña, N. Mex, (new)	548
Fort Wingate, Santa Aña, N. Mex., (old)	496
Greenhorn, Huerfano, Colo	126
Julesburgh, Weld, Colo	442
Kansas City, Jackson, Mo	539
Kit Carson, Greenwood, Colo	50
Laramie City, Albany, Wyo	359
Leavenworth City, Leavenworth, Kans	532
North Platte, Lincoln, Nebr	528

FORT LYON, COLO., TO—

	Miles.
Ogden City, Weber, Utah	816
Omaha City, Douglas, Nebr	819
Pueblo, Pueblo, Colo	98
Salt Lake City, Salt Lake, Utah	853
Santa Fé, Santa Fé, N. Mex	340
South Side, Bent, Colo	56
Trinidad, Las Animas, Colo	118
Washington, D. C	1,728

FORT McHENRY, MD., TO—

Baltimore, Baltimore, Md	3
Boston, Suffolk, Mass	429
Fort Monroe, Elizabeth City, Va	188
Fort Preble, Cumberland, Me	538
Fort Sullivan, Washington, Me	798
Fort Trumbull New London, Conn	319
Fort Wadsworth, Richmond, N. Y	202
Fort Warren, Suffolk, Mass	436
Fort Washington, Prince George's, Md	56
Fort Wood, New York, N. Y	195
New London, New London, Conn	319
New York, New York, N. Y	193
Philadelphia, Philadelphia, Pa	104
Pikesville Arsenal, Baltimore, Md	11
Plattsburgh Barracks, Clinton, N. Y	506
Raleigh, Wake, N. C	357
Rutherfordton, Rutherford, N. C	535
Washington, D. C	41
Watertown Arsenal, Middlesex, Mass	437

FORT McINTOSH, (LAREDO P. O.,) TEX., TO—

Fort McKavett, Menard, Tex	340
Fort Quitman, El Paso, Tex	784
Fort Richardson, Jack, Tex	496
Fort Stockton, Pecos, Tex	564
Galveston, Galveston, Tex	423
Houston, Harris, Tex	474
Indianola, Calhoun, Tex	293
Marshall, Harrison, Tex	548
New Orleans, Orleans, La	725
Orange, Orange, Tex	584
Ringgold Barracks, Starr, Tex	141
Rio Grande City, Starr, Tex	140
San Antonio, Bexar, Tex	165
Victoria, Victoria, Tex	339

FORT McKAVETT, TEX., TO—

	Miles.
Dallas, Dallas, Tex	375
Denison, Tex	375
Fort Quitman, El Paso, Tex	445
Fort Richardson, Jack, Tex	275
Fort Sill, Choctaw N., Ind. Ter	571
Fort Smith, Sebastian, Ark	679
Fort Stockton, Pecos, Tex	225
Fredericksburgh, Gillespie, Tex	104
Galveston, Galveston, Tex	448
Houston, Harris, Tex	399
Indianola, Calhoun, Tex	333
Laredo, Webb, Tex	340
Mason, Mason, Tex	59
Menardville, Menard, Tex	21
New Orleans, Orleans, La	750
Orange, Orange, Tex	509
Ringgold Barracks, Starr, Tex	481
Rio Grande City, Starr, Tex	480
San Antonio, Bexar, Tex	175
Victoria, Victoria, Tex	287
Waco, McLennan, Tex	286
Washington, D. C	1,946

FORT McKEEN, DAK.

See Fort Abraham Lincoln.

FORT McPHERSON, NEBR., TO—

Fort Reynolds, Pueblo, Colo	489
Fort Sanders, Albany, Wyo	299
Fort Sedgwick, Weld, Colo	109
Fort Union, Mora, N. Mex	734
Julesburgh, Weld, Colo	106
Laramie City, Albany, Wyo	301
Leavenworth City, Leavenworth, Kans	463
McPherson, Lincoln, Nebr	6
North Platte, Lincoln, Nebr	20
Ogden City, Weber, Utah	758
Omaha City, Douglas, Nebr	283
Pueblo, Pueblo, Colo	469
Salt Lake City, Salt Lake, Utah	795
Sidney, Cheyenne, Nebr	143
Washington, D. C	1,549

FORT McRAE, N. MEX., TO—

Fort Reynolds, Pueblo, Colo	612
Fort Selden, Doña Aña, N. Mex	58
Fort Stanton, Socorro, N. Mex	318
Fort Sumner, San Miguel, N. Mex	346

FORT McRAE, N. MEX., TO—

	Miles.
Fort Union, Mora, N. Mex	327
Fort Wingate, Santa Aña, N. Mex., (new)	259
Grenada, Colo	590
Kit Carson, Greenwood, Colo	611
Santa Fé, Santa Fé, N. Mex	221
Washington, D. C	2,341

FORT MACON, N. C., TO—

Aquia Creek, Stafford, Va	338
Baltimore, Baltimore, Md	432
Beaufort, Carteret, N. C	2
Boston, Suffolk, Mass	758
Fort McHenry, Baltimore, Md	435
Fort Monroe, Elizabeth City, Va	269
Fort Preble, Cumberland, Me	867
Fort Sullivan, Washington, Me	1,127
Fort Trumbull, New London, Conn	748
Fort Wadsworth, Richmond, N. Y	631
Fort Warren, Suffolk, Mass	765
Fort Washington, Prince George's, Md	379
Fort Wood, New York, N. Y	624
Goldsborough, Wayne, N. C	97
Morehead City, Carteret, N. C	2
New London, New London, Conn	748
New York, New York, N. Y	622
Philadelphia, Philadelphia, Pa	533
Pikesville Arsenal, Baltimore, Md	440
Plattsburgh Barracks, Clinton, N. Y	935
Raleigh, Wake, N. C	145
Richmond, Henrico, Va	263
Rutherfordton, Rutherford, N. C	400
Washington, D. C	394
Watertown Arsenal, Middlesex, Mass	766
Willett's Point, Queens, N. Y	642
Wilmington, New Hanover, N. C	179

FORTRESS MONROE, VA., TO—

Baltimore, Baltimore, Md	185
Boston, Suffolk, Mass	611
Fort Preble, Cumberland, Me	720
Fort Sullivan, Washington, Me	980
Fort Trumbull, New London, Conn	501
Fort Wadsworth, Richmond, N. Y	384
Fort Warren, Suffolk, Mass	618
Fort Washington, Prince George's, Md	238
Fort Wood, New York, N. Y	377
New London, New London, Conn	501
New York, New York, N. Y	375
Norfolk, Norfolk, Va	15
Philadelphia, Philadelphia, Pa	286
Pikesville Arsenal, Baltimore, Md	193
Plattsburgh Barracks, Clinton, N. Y	688

FORTRESS MONROE, VA., TO—

	Miles.
Raleigh, Wake, N. C	193
Rutherfordton, Rutherford, N. C	448
Washington, D. C	223
Watertown Arsenal, Middlesex, Mass	619
Willett's Point, Queens, N. Y	395
Wilmington, New Hanover, N. C	256

FORT NIAGARA, N. Y., TO—

Buffalo, Erie, N. Y	35
Detroit, Wayne, Mich	245
Lewiston, Niagara, N. Y	7
Madison Barracks, Jefferson, N. Y	266½
Suspension Bridge, Niagara, N. Y	13
Washington, D. C	483
Youngstown, Niagara, N. Y	1

FORT ONTARIO, N. Y.,

At Oswego City, which see.

FORT PEMBINA, DAK., TO—

Fort Randall, Todd, Dak	832
Fort Ransom, Ransom, Dak	254
Fort Ripley, Morrison, Minn	406
Fort Snelling, Hennepin, Minn	454
Fort Sully, Sully, Dak	1,017
Fort Thompson, Buffalo, Dak	933
Fort Wadsworth, Stone, Dak	262
Georgetown, Clay, Minn	134
Holy Cross, Clay, Minn	160
Lake Winnepeg, Brit. Amer	114
Minneapolis, Hennepin, Minn	456
Moorhead, Clay, Minn	153
Pembina, Pembina, Dak	3
Ponka Agency, Todd, Dak	818
Saint Cloud, Stearns, Minn	372
Saint Joseph, Pembina, Dak	33
Saint Paul, Ramsey, Minn	447
Sioux City, Woodbury, Iowa	688
Stone Fort, Brit. Amer	91
Vermillion, Clay, Dak	734
Washington, D. C	1,662
Whetstone Agency, Gregory, Dak	848
Yankton, Yankton, Dak	756

FORT PORTER, N. Y.,

Is within the city limits of Buffalo, N. Y.

FORT PREBLE, ME., TO—

	Miles.
Baltimore, Baltimore, Md	535
Boston, Suffolk, Mass	109
Fort Sullivan, Washington, Me	262
Fort Trumbull, New London, Conn	220
Fort Wadsworth, Richmond, N. Y	354
Fort Warren, Suffolk, Mass	116
Fort Washington, Prince George's, Md	588
Fort Wood, New York, N. Y	347
New London, New London, Conn	219
New York, New York, N. Y	345
Philadelphia, Philadelphia, Pa	434
Pikesville Arsenal, Baltimore, Md	543
Plattsburgh Barracks, Clinton, N. Y	283
Raleigh, Wake, N. C	889
Rutherfordton, Rutherford, N. C	936
Washington, D. C	573
Watertown Arsenal, Middlesex, Mass	117
Willett's, Point, Queens, N. Y	365
Wilmington, New Hanover, N. C	954

FORT PULASKI, GA., TO—

Frankfort, Franklin, Ky	857
Humboldt, Gibson, Tenn	747
Huntsville, Madison, Ala	554
Jackson, Hinds, Miss	656
Key West, Monroe, Fla	712
Lebanon, Marion, Ky	799
Louisville, Jefferson, Ky	792
Meridian, Lauderdale, Miss	560
Mobile, Mobile, Ala	588
Mt. Sterling, Montgomery, Ky	920
Nashville, Davidson, Tenn	607
Paducah, McCracken, Ky	779
Savannah, Chatham, Ga	15
Shelbyville, Shelby, Ky	823
St. Augustine, St. John's, Fla	332
Washington, D. C	697

FORT QUITMAN, TEX., TO—

Dallas, Dallas, Tex	710
Denison, Tex	710
Fort Richardson, (Jacksborough P. O.,) Jack, Tex	610
Fort Sill, Choctaw N., Ind. Ter	906
Fort Stockton, Pecos, Tex	220
Galveston, Galveston, Tex	892
Houston, Harris, Tex	843
Indianola, Calhoun, Tex	777
Laredo, (Ft. McIntosh,) Webb, Tex	784
New Orleans, Orleans, La	1,194
Ringgold Barracks, Starr, Tex	773
Rio Grande City, Starr, Tex	772
San Antonio, Bexar, Tex	619
Victoria, Victoria, Tex	731
Waco, McLennan, Tex	731
Washington, D. C	2,205

FORT RANDALL, DAK., TO—

	Miles.
Fort Ransom, Ransom, Dak	720
Fort Rice, Morton, Dak	443
Fort Ripley, Morrison, Minn	532
Fort Snelling, Hennepin, Minn	402
Fort Stevenson, Stevens, Dak	566
Fort Sully, Sully, Dak	185
Fort Thompson, Buffalo, Dak	101
Fort Totten, Ramsey, Dak	694
Grand River, Boreman, Dak	278
Greenwood, Charles Mix, Dak	25
McCauleyville, Wilkins, Minn	645
Omaha City, Douglas, Nebr	235
Ponka Agency, Todd, Dak	46
Saint Paul, Ramsey, Minn	407
Sioux City, Woodbury, Iowa	132
Vermillion, Clay, Dak	98
Washington, D. C	1,449
Whetstone Agency, Gregory, Dak	16
White Swan, Charles Mix, Dak	½
Yankton, Yankton, Dak	71

FORT RANSOM, DAK., TO—

Cheyenne River Depot	30
Fort Ripley, Morrison, Minn	286
Fort Seward, Dak	41
Fort Snelling, Hennepin, Minn	308
Fort Stevenson, Stevens, Dak	258
Fort Totten, Ramsey, Dak	121
Georgetown, Clay, Minn	120
Mouse River, Stevens, Dak	205
Pembina, Pembina, Dak	257
Saint Cloud, Stearns, Minn	238
Saint Joseph, Pembina, Dak	287
Saint Paul, Ramsey, Minn	307
Sauk Centre, Stearns, Minn	178
Sioux City, Woodbury, Iowa	575
Stone Fort, Brit. Amer	345
Washington, D. C	1,522
Willmar, Kandiyohi, Minn	202
Yankton, Yankton, Dak	643

FORT REYNOLDS, COLO., TO—

Badito, Huerfano, Colo	80
Bent's Fort, Bent, Colo	60
Fort Sanders, Albany, Wyo	298
Fort Sedgwick, Weld, Colo	386
Fort Selden, Doña Aña, N. Mex	621
Fort Sumner, San Miguel, N. Mex	374
Fort Union, Mora, N. Mex	226
Fort Wallace, Wallace, Kans	193
Fort Wingate, Santa Aña, N. Mex., (new)	540

FORT REYNOLDS, COLO., TO—

	Miles.
Fort Wingate, Santa Aña, N. Mex., (old)	488
Greenhorn, Huerfano, Colo	48
Julesburgh, Weld, Colo	383
Kansas City, Jackson, Mo	617
Kit Carson, Greenwood, Colo	128
Laramie City, Albany, Wyo	300
Leavenworth City, Leavenworth, Kans	608
North Platte, Lincoln, Nebr	469
Ogden City, Weber, Utah	757
Omaha City, Douglas, Nebr	760
Pueblo, Pueblo, Colo	20
Salt Lake City, Salt Lake, Utah	794
Santa Fé, Santa Fé, N. Mex	332
South Side, Bent, Colo	22
Trinidad, Las Animas, Colo	117
Washington, D. C	1,806

FORT RICE, DAK., TO—

Fort Ripley, Morrison, Minn	677
Fort Seward, Dak	130
Fort Snelling, Hennepin, Minn	697
Fort Stevenson, Stevens, Dak	120
Fort Sully, Sully, Dak	258
Fort Thompson, Buffalo, Dak	343
Fort Totten, Ramsey, Dak	210
Grand River Agency, Boreman, Dak	165
Omaha City, Douglas, Nebr	678
Ponka Agency, Todd, Dak	489
Saint Paul, Ramsey, Minn	505
Sioux City, Woodbury, Iowa	574
Vermillion, Clay, Dak	541
Washington, D. C	1,864
Yankton, Yankton, Dak	514

FORT RICHARDSON, TEX., TO—

Dallas, Dallas, Tex	100
Denison, Tex	100
Fort Sill, Choctaw N., Ind. Ter	296
Fort Stockton, Pecos, Tex	390
Fort Worth, Tarrant, Tex	71
Galveston, Galveston, Tex	399
Hempstead, Austin, Tex	298
Hillsborough, Hill, Tex	122
Houston, Harris, Tex	348
Indianola, Calhoun, Tex	501
Jacksborough, Jack, Tex	1
Laredo, Webb, Tex	496
Longview, Upshur, Tex	258
Marlin, Falls, Tex	187
New Orleans, Orleans, La	701
Orange, Orange, Tex	458

FORT RICHARDSON, TEX., TO—

	Miles.
Ringgold Barracks, Starr, Tex	637
Rio Grande City, Starr, Tex	636
Saint Louis, St. Louis, Mo	686
San Antonio, Bexar, Tex	331
Sherman, Grayson, Tex	103
Vicksburgh, Warren, Miss	518
Victoria, Victoria, Tex	390
Vinita, Ind. Ter	321
Waco, McLennan, Tex	167
Washington, D. C	1,592
Weatherford, Parker, Tex	41

FORT RIPLEY, MINN., TO—

Fort Snelling, Hennepin, Minn	125
Fort Sully, Buffalo, Dak	717
Fort Totten, Ramsey, Dak	385
Pembina, Pembina, Dak	409
Saint Cloud, Stearns, Minn	50
Saint Paul, Ramsey, Minn	125
Washington, D. C	1,340

FORT SANDERS, WYO., TO—

Fort Sedgwick, Weld, Colo	196
Fort Union, Mora, N. Mex	543
Fort Wallace, Wallace, Kans	372
Julesburgh, Weld, Colo	193
Kit Carson, Greenwood, Colo	307
Laramie City, Albany, Wyo	3
Leavenworth City, Leavenworth, Kans	750
North Platte, Lincoln, Nebr	279
Ogden City, Weber, Utah	459
Omaha City, Douglas, Nebr	570
Pueblo, Pueblo, Colo	278
Salt Lake City, Salt Lake, Utah	496
Santa Fé, Santa Fé, N. Mex	649
Washington, D. C	1,836

FORT SEDGWICK, COLO., TO—

Fort Union, Mora, N. Mex	631
Fort Wallace, Wallace, Kans	460
Julesburgh, Weld, Colo	3
Kit Carson, Greenwood, Colo	395
Laramie City, Albany, Wyo	198
Leavenworth City, Leavenworth, Kans	560
North Platte, Lincoln, Nebr	89
Ogden City, Weber, Utah	655
Omaha City, Douglas, Nebr	380

FORT SEDGWICK, COLO., TO—

	Miles.
Pueblo, Pueblo, Colo	366
Salt Lake City, Salt Lake, Utah	692
Santa Fé, Santa Fé, N. Mex	737
Sidney, Cheyenne, Nebr	40
Trinidad, Las Animas, Colo	463
Washington, D. C	1,646

FORT SELDEN, N. MEX., TO—

Doña Aña, Doña Aña, N. Mex	12
Fort Stanton, Socorro, N. Mex	386
Fort Sumner, San Miguel, N. Mex	414
Fort Union, Mora, N. Mex	395
Fort Wallace, Wallace, Kans	798
Fort Wingate, Santa Aña, N. Mex	355
Grenada, Colo	658
Las Cruces, Doña Aña, N. Mex	18
Leasburgh, Doña Aña, N. Mex	1
Mesilla, Doña Aña, N. Mex	21
Ralston, Grant, N. Mex	159
Rio Miembres, Grant, N. Mex	101
Santa Fé, Santa Fé, N. Mex	289
Washington, D. C	2,409

FORT SEWARD, DAK., TO—

Fort Snelling, Minn	363
Fort Stevenson, Dak	195
Fort Sully, Dak	388
Fort Totten, Dak	83
Mendota, Minn	365
Minneapolis, Hennepin, Minn	350
Moorhead, Minn	98
St. Paul, Minn	375
Sioux City, Iowa	704
Washington, D. C	2,022
Whetstone Agency, Dak	489
Yankton, Dak	644

FORT SHAW, MONT., TO—

Deer Lodge City, Deer Lodge, Mont	135
Gallatin City, Gallatin, Mont	148
Helena, Lewis and Clarke, Mont	80
Missoula, Missoula, Mont	220
Spokan Bridge, Stevens, Wash	490
Virginia City, Madison, Mont	203
Walla Walla, Walla Walla, Wash	645
Washington, D. C	2,848

FORT SILL, IND. TER., TO—

	Miles.
Caddo, Ind. Ter	165
Denison, Tex	196
Fort Smith, Sebastian, Ark	269
Fort Stockton, Pecos, Tex	675
Fort Washita, Chickasaw N., Ind. Ter	167
Galveston, Galveston, Tex	582
Houston, Harris, Tex	532
Jacksborough, Jack, Tex	296
Leavenworth City, Leavenworth, Kans	672
Saint Louis, Saint Louis, Mo	720
San Antonio, Bexar, Tex	674
Vinita, Cherokee N., Ind. Ter	355
Washington, D. C	1,626

FORT SMITH, ARK., TO—

Bentonville, Benton, Ark	100
Boggy Depot, Choctaw N., Ind. Ter	129
Dardanelle, Yell, Ark	95
Fort Harker, Ellsworth, Kans	429
Galveston, Galveston, Tex	587
Houston, Harris, Tex	536
Junction City, Davis, Kans	349
Little Rock, Pulaski, Ark	206
Memphis, Shelby, Tenn	341
Neosho, Newton, Mo	147
New Orleans, Orleans, La	734
Saint Louis, Saint Louis, Mo	461
Vicksburgh, Warren, Miss	396
Washington, D. C	1,367

FORT SNELLING, MINN., TO—

Fort Stevenson, Stevens, Dak	558
Fort Sully, Sully, Dak	588
Fort Thompson, Buffalo, Dak	503
Fort Totten, Ramsey, Dak	398
Mendota, Dakota, Minn	2
Minneapolis, Hennepin, Minn	13
Moorhead, Clay, Minn	265
Pembina, Pembina, Dak	422
Saint Paul, Ramsey, Minn	7
Sioux City, Woodbury, Iowa	268
Washington, D. C	1,222

FORT STANTON, N. MEX., TO—

Fort Sumner, San Miguel, N. Mex	99
Fort Union, Mora, N. Mex	345
Fort Wingate, Santa Aña, N. Mex	305
Mesilla, Doña Aña, N. Mex	407
Rio Miembres, Grant, N. Mex	487
Santa Fé, Santa Fé, N. Mex	239
Washington, D. C	2,359

FORT STEILACOOM, WASH., TO—

	Miles.
Fort Stevens, Clatsop, Oreg	175
Jacksonville, Jackson, Oreg	459
Lewistown, Nez Percés, Idaho	464
Monticello, Cowlitz, Wash	111
Olympia, Thurston, Wash	26
Portland, Multnomah, Oreg	164
Red Bluff, Tehama, Cal	646
Reno, Washoe, Nev	896
Sacramento, Sacramento, Cal	784
Sacramento Junction, Placer, Cal	766
San Francisco, San Francisco, Cal	809
Umatilla, Umatilla, Oreg	324
Vancouver, Clarke, Wash	182
Walla Walla, Walla Wall, Wash	379
Wallula, Walla Walla, Wash	349
Washington, D. C	3,244

FORT STEVENS, OREG., TO—

Jacksonville, Jackson, Oreg	397
Lewiston, Nez Percés, Wash	464
Monticello, Cowlitz, Wash	64
Olympia, Thurston, Wash	149
Portland, Multnomah, Oreg	104
Red Bluff, Tehamah, Cal	584
Reno, Washoe, Nev	840
Sacramento, Sacramento, Cal	722
San Francisco, San Francisco, Cal	650
Sitka, Alaska	983
Umatilla, Umatilla, Oreg	324
Vancouver, Clarke, Wash	122
Victoria, Brit. Col	291
Walla Walla, Walla Walla, Wash	349
Washington, D. C	3,231

FORT STEVENSON, DAK., TO—

Fort Sully, Sully, Dak	381
Fort Thompson, Buffalo, Dak	466
Fort Totten, Ramsey, Dak	126
Grand River Agency, Boreman, Dak	288
Mendota, Dakota, Minn	560
Omaha City, Douglas, Nebr	801
Saint Paul, Ramsey, Minn	568
Sioux City, Woodbury, Iowa	697
Vermillion, Clay, Dak	666
Washington, D. C	1,758
Whetstone Agency, Gregory, Dak	584
Yankton, Yankton, Dak	639

FORT STOCKTON, TEX., TO—

	Miles.
Chihuahua, Mex	320
Denison, Tex	490
Galveston, Galveston, Tex	672
Houston, Harris, Tex	623
Indianola, Calhoun, Tex	557
New Orleans, Orleans, La	974
Presidio del Norte, Mex	140
Ringgold Barracks, Starr, Tex	705
Rio Grande City, Starr, Tex	704
San Antonio, Bexar, Tex	399
Victoria, Victoria, Tex	511
Waco, McLennan, Tex	511
Washington, D. C	1,982

FORT SULLIVAN, (EASTPORT,) ME., TO—

Baltimore, Baltimore, Md	795
Boston, Suffolk, Mass	369
Calais, Washington, Me	29
Fort Trumbull, New London, Conn	380
Fort Wadsworth, Richmond, N. Y	614
Fort Warren, Suffolk, Mass	376
Fort Washington, Prince George's, Md	848
Fort Wood, New York, N. Y	607
New London, New London, Conn	369
New York, New York, N. Y	605
Philadelphia, Philadelphia, Pa	694
Pikesville Arsenal, Baltimore, Md	803
Plattsburgh Barracks, (Plattsburgh P. O.,) Clinton, N. Y	500
Raleigh, Wake, N. C	1,149
Rutherfordton, Rutherford, N. C	1,327
Washington, D. C	833
Watertown Arsenal, Middlesex, Mass	477
Willett's Point, Queens, N. Y	625
Wilmington, New Hanover, N. C	1,214

FORT SULLY, DAK., TO—

Fort Thompson, Buffalo, Dak	85
Fremont, Dodge, Nebr	429
Grand River, Boreman, Dak	125
Omaha, Douglas, Nebr	416
Ponka Agency, Todd, Dak	231
Saint Paul Ramsey, Minn	588
Sioux City, Woodbury, Iowa	316
Vermillion, Clay, Dak	283
Washington, D. C	1,628
Whetstone Agency, Gregory, Dak	171
White Swan, Charles Mix, Dak	186
Yankton, Yankton, Dak	256

FORT THOMPSON, DAK., TO—

	Miles.
Fort Totten, Ramsey, Dak	594
Grand River, Boreman, Dak	178
Omaha City, Douglas, Nebr	336
Ponka Agency, Todd, Dak	147
Saint Paul, Ramsey, Minn	747
Sioux City, Woodbury, Iowa	236
Vermillion, Clay, Dak	199
Washington, D. C	1,525
Whetstone Agency, Gregory, Dak	86
Yankton, Yankton, Dak	172

FORT TOTTEN, DAK., TO—

Fort Wadsworth, Stone, Dak	240
Georgetown, Clay, Dak	158
Minneapolis, Hennepin, Minn	385
Pembina, Pembina, Dak	140
Pomme de Terre, Grant, Minn	226
Saint Paul, Ramsey, Minn	396
Sioux City, Woodbury, Iowa	668
Stone Fort, Brit. Amer	218
Walhalla, Pembina, Dak	109
Washington, D. C	1,592
Yankton, Yankton, Dak	728

FORT TRUMBULL, CONN., TO—

Baltimore, Baltimore, Md	316
Boston, Suffolk, Mass	110
Fort Wadsworth, Richmond, N. Y	135
Fort Warren, Suffolk, Mass	117
Fort Washington, Prince George's, Md	369
Fort Wood, N. Y. Harbor	28
New London, New London, Conn	½
New York, New York, N. Y	126
Philadelphia, Philadelphia, Pa	215
Pikesville Arsenal, Baltimore, Md	324
Plattsburgh Barracks, Clinton, N. Y	288
Raleigh, Wake, N. C	670
Rutherfordton, Rutherford, N. C	848
Washington, D. C	354
Watertown Arsenal, Middlesex, Mass	118
Willett's Point, Queens, N. Y	148
Wilmington, New Hanover, N. C	735

FORT TULEROSA, N. MEX., TO—

Camp Apache, Ariz	164
Fort Union, Mora, N. Mex	322
Grenada, Colo	585
Kansas City, Jackson, Mo	1,085
Pueblo, Pueblo, Colo	528
Sargent, Kans	598
Washington, Washington, D. C	2,274

FORT UNION, N. MEX., TO—

	Miles.
Cimarron, Mora, N. Mex	51
Fort Wingate, Sañta Aña, N. Mex., (new)	270
Greenhorn, Huerfano, Colo	177
Grenada, Colo	263
Kansas City, Jackson, Mo	763
Kit Carson, Greenwood, Colo	284
La Junta, Mora, N. Mex	9
Las Vegas, Doña Aña, N. Mex	30
Leavenworth City, Leavenworth, Kans	761
Mesilla, Doña Aña, N. Mex	416
Pueblo, Pueblo, Colo	213
Santa Fé, Santa Fé, N. Mex	106
Tecolote, San Miguel, N. Mex	42
Trinidad, Las Animas, Colo	116
Washington, D. C	1,952

FORT WADSWORTH, DAK., TO—

Georgetown, Clay, Minn	127
Lake Winnepeg, Brit. Amer	367
Saint Paul, Ramsey, Minn	306
Washington, D. C	1,521

FORT WADSWORTH, N. Y., TO—

Baltimore, Baltimore, Md	199
Boston, Suffolk, Mass	245
Fort Warren, Suffolk, Mass	252
Fort Washington, Prince George's, Md	252
Fort Wood, N. Y. Harbor	11
New London, New London, Conn	135
New York, New York, N. Y	9
Philadelphia, Philadelphia, Pa	98
Pikesville Arsenal, Baltimore, Md	207
Plattsburgh Barracks, Clinton, N. Y	322
Raleigh, Wake, N. C	553
Rutherfordton, Rutherford, N. C	731
Washington, D. C	237
Watertown Arsenal, Middlesex, Mass	253
Willett's Point, Queens, N. Y	29
Wilmington, New Hanover, N. C	618

FORT WALLACE, KANS., TO—

Kansas City, Jackson, Mo	424
Kit Carson, Greenwood, Colo	65
Laramie City, Albany, Wyo	374
Lawrence, Douglas, Kans	384
Leavenworth, Leavenworth, Kans	417
Pueblo, Pueblo, Colo	213
Salt Lake City, Salt Lake, Utah	868
Santa Fé, Santa Fé, N. Mex	507
Trinidad, Las Animas, Colo	233
Washington, D. C	1,613

FORT WARREN, MASS., TO—

	Miles.
Baltimore, Baltimore, Md	433
Boston, Suffolk, Mass	7
Fort Washington, Prince George's, Md	486
Fort Wood, N. Y. Harbor	245
New London, New London, Conn	117
Newport, Newport, R. I	76
New York, New York, N. Y	243
Philadelphia, Philadelphia, Pa	332
Pikesville Arsenal, Baltimore, Md	441
Plattsburgh Barracks, Clinton, N. Y	263
Raleigh, Wake, N. C	787
Rutherfordton, Rutherford, N. C	965
Washington, D. C	471
Watertown Arsenal, Middlesex, Mass	15
Willett's Point, Queens, N. Y	263
Wilmington, New Hanover, N. C	852

FORT WASHINGTON, MD., TO—

Alexandria, Alexandria, Va	9
Baltimore, Baltimore, Md	53
Boston, Suffolk, Mass	479
Fort Wood, N. Y. Harbor	245
New York, New York, N. Y	243
Philadelphia, Philadelphia, Pa	154
Pikesville Arsenal, Baltimore, Md	61
Plattsburgh Barracks, Clinton, N. Y	571
Raleigh, Wake, N. C	301
Rutherfordton, Rutherford, N. C	479
Washington, D. C	15
Watertown Arsenal, Middlesex, Mass	487
Willett's Point, Queens, N. Y	263
Wilmington, New Hanover, N. C	366

FORT WHIPPLE, (PRESCOTT,) ARIZ., TO—

Fort Yuma, San Diego, Cal	318
La Paz, Yuma, Ariz	211
Los Angeles, Los Angeles, Cal	483
Maricopa Wells, Pima, Ariz	186
Sacramento, Sacramento, Cal	902
San Bernardino, San Bernardino, Cal	421
San Diego, San Diego, Cal	515
San Francisco, San Francisco, Cal	930
Tucson, Pima, Ariz	311
Washington, D. C	3,155
Wickenburgh, Yavapai, Ariz	91
Wilmington, Los Angeles, Cal	503

FORT WINGATE, N. MEX., (NEW,) TO—

	Miles.
Fort Wingate, Sañta Aña, N. Mex., (old)	52
Grenada, Colo	537
Kit Carson, Greenwood, Colo	652
Pueblo, Pueblo, Colo	520
Santa Fé, Santa Fé, N. Mex	208
Trinidad, Las Animas, Colo	430
Washington, D. C	2,266
Zuni, Valencia, N. Mex	35

FORT WOOD, N. Y., TO—

Baltimore, Baltimore, Md	192
Boston, Suffolk, Mass	238
New York, New York, N. Y	2
Philadelphia, Philadelphia, Pa	91
Pikesville Arsenal, Baltimore, Md	200
Plattsburgh Barracks, Clinton, N. Y	315
Raleigh, Wake, N. C	546
Rutherfordton, Rutherford, N. C	724
Washington, D. C	230
Watertown Arsenel, Middlesex, Mass	246
Willett's Point, Queens, N. Y	22
Wilmington, New Hanover, N. C	611

FORT YUMA, CAL., TO—

Camp Apache, Maricopa, Ariz	519
Camp Bowie, Pima, Ariz	437
Camp Crittenden, Pima, Ariz	376
Camp Date Creek, Yavapai, Ariz	254
Camp Grant, Pima, Ariz	291
Camp Hualpai, Yavapai, Ariz	353
Camp Lowell, (Tucson,) Pima, Ariz	325
Camp McDowell, Maricopa, Ariz	264
Camp Mohave, Mohave, Ariz	240
Camp Verde, Yavapai, Ariz	445
Fort Whipple, Yavapai, Ariz	318
La Paz, Yuma, Ariz	107
Los Angeles, Los Angeles, Cal	343
Maricopa, Wells, Pima, Ariz	200
Prescott, Yavapai, Ariz	318
San Antonia, Bexar, Tex	1,379
San Barnardino, San Barnardino, Cal	317
San Diego, San Diego, Cal	197
San Francisco, San Francisco, Cal	714
Santa Fé, Santa Fé, N. Mex	935
Tucson, Pima, Ariz	325
Washington, D. C	3,838
Wickenburgh, Yavapai, Ariz	247
Wilmington, Los Angles, Cal	363

FRANKFORT, KY., TO—

	Miles.
Humboldt, Tenn	318
Huntsville, Ala	396
Jackson, Miss	593
Lebanon, Ky	67
Louisville, Ky	65
Mount Sterling, Ky	61
Nashville, Tenn	250
Paducah, Ky	340
Washington, D. C	683

FREDERICKSBURGH, TEX., TO—

Boerne, Kendall, Tex	42
Cherry Spring, Gillespie, Tex	20
Comfort, Kerr, Tex	26
Galveston, Galveston, Tex	344
Houston, Harris, Tex	295
Laredo, Webb, Tex	236
Leon Springs, Bexar, Tex	53
Mason, Mason, Tex	45
Menardville, Menard, Tex	83
New Orleans, Orleans, La	646
Ringgold Barracks, Starr, Tex	377
Rio Grande City, Starr, Tex	376
San Antonio, Bexar, Tex	71
Victoria, Victoria, Tex	183
Waco, McLennan, Tex	182
Washington, D.C	1,842

FREMONT, NEB., TO—

Kelton, Box Elder, Utah	1,074
Leavenworth, Leavenworth, Kans	227
Missouri Valley Junction, Harrison, Iowa	38
Ogden City, Weber, Utah	982
Omaha City, Douglas, Nebr	47
Reno, Washoe, Nev	1,571
Sacramento, Sacramento, Cal	1,725
Sacramento Junction, Placer, Cal	1,707
San Francisco, San Francisco, Cal	1,808
Sioux City, Woodbury, Iowa	113
Washington, D. C	1,313

GALLATIN, MONT., TO—

Gaffney, Madison, Mont	138
Helena, Lewis and Clarke, Mont	68
New York, N. Y	2,929
Omaha City, Douglas, Nebr	1,482
Spokan Bridge, Stevens, Wash	478
Virginia City, Madison, Mont	110
Washington, D. C	2,800

GALVESTON, TEX., TO

	Miles.
Brashear City, St. Mary's, La	220
Crockett, Houston, Tex	211
Denison, Tex	386
Hallsville, Harrison, Tex	321
Harrisburgh, Harris, Tex	47
Houston, Harris, Tex	51
Indianola, Calhoun, Tex	130
Liberty, Liberty, Tex	92
Marshall, Harrison, Tex	333
Navasota, Grimes, Tex	121
New Orleans, Orleans, La	302
Orange, Orange, Tex	161
Ringgold Barracks, (Rio Grande City,) Starr, Tex	404
Rio Grande City, Starr, Tex	403
San Antonio, Bexar, Tex	273
Victoria, Victoria, Tex	176
Waco, McLennan, Tex	242
Washington, D. C	1,498

GEORGETOWN, MINN., TO—

Mendota, Dakota, Minn	287
Minneapolis, Hennepin, Minn	280
Pembina, Pembina, Dak	137
Pomme de Terre, Grant, Minn	113
Saint Cloud, Stearns, Minn	222
Saint Joseph, Pembina, Dak	167
Saint Paul, Ramsey, Minn	291
Sioux City, Woodbury, Iowa	563
Stone Fort, Brit. Amer	224
Washington, D. C	1,506
Willmar, Kandiyohi, Minn	186
Yankton, Yankton, Dak	627

GRAND RIVER AGENCY, DAK., TO—

Fort Rice, Morton, Dak	165
Fort Sully, Sully, Dak	93
Omaha, Douglas, Nebr	533
Saint Paul, Ramsey, Minn	705
Sioux City, Woodbury, Iowa	433
Washington, D. C	1,747
Yankton, Yankton, Dak	367

GUAYMAS, MEX., TO—

Camp Wallen, Pima, Ariz	317
Tucson, Pima, Ariz	351

HALLSVILLE, TEX., TO—

	Miles.
Harrisburgh, Harris, Tex	274
Houston, Harris, Tex	270
Indianola, Calhoun, Tex	413
Longview, Upshur, Tex	10
Marshall, Harrison, Tex	14
Navasota, Grimes, Tex	200
New Orleans, Orleans, La	478
Ringgold Barracks, Starr, Tex	745
Rio Grande City, Starr, Tex	744
San Antonio, Bexar, Tex	369
Sherman, Grayson, Tex	192
Shreveport, Caddo, La	56
Vicksburgh, Warren, Miss	250
Victoria, Victoria, Tex	367
Washington, D. C	1,311

HARRISBURGH, TEX., TO—

Houston, Harris, Tex	6
New Orleans, Orleans, La	347
Orange, Orange, Tex	116
San Antonio, Bexar, Tex	224
Waco, McLennan, Tex	197
Washington, D. C	1,543

HELENA, MONT., TO—

Deer Lodge City, Deer Lodge, Mont	55
Missoula, Missoula, Mont	140
New York, N. Y	2,860
Omaha City, Douglas, Nebr	1,482
Spokan Bridge, Stevens, Wash	410
Virginia City, Madison, Mont	123
Walla Walla, Walla Walla, Wash	565
Washington, D. C	2,732

HILLSBOROUGH, TEX., TO—

Hempstead, Austin, Tex	176
Houston, Harris, Tex	226
Indianola, Calhoun, Tex	314
Laredo, Webb, Tex	375
New Orleans, Orleans, La	579
Orange, Orange, Tex	336
Ringgold Barracks, Starr, Tex	681
Rio Grande City, Starr, Tex	680
San Antonio, Bexar, Tex	210
Victoria, Victoria, Tex	269
Waco, McLennan, Tex	35
Washington, D. C	1,598

HOUSTON, TEX., TO—

	Miles.
Indianola, Calhoun, Tex	181
Laredo, Webb, Tex	474
Liberty, Liberty, Tex	41
Marlin, Falls, Tex	161
Marshall, Harrison, Tex	70
Orange, Orange, Tex	110
Ringgold Barracks, Starr, Tex	455
Rio Grande City, Starr, Tex	454
San Antonio, Bexar, Tex	224
Victoria, Victoria, Tex	163
Waco, McLennan, Tex	181
Washington, D. C	1,549

HUMBOLDT, TENN., TO—

Huntsville, Madison, Ala	192
Jackson, Hinds, Miss	275
Key West, Monroe, Fla	1,059
Lebanon, Marion, Ky	260
Louisville, Jefferson, Ky	296
Meridian, Lauderdale, Miss	267
Mobile, Mobile, Ala	402
Mount Sterling, Montgomery, Ky	379
Nashville, Davidson, Tenn	150
Paducah, McCracken, Ky	102
Savannah, Chatham, Ga	732
Shelbyville, Shelby, Ky	284
San Augustine, St. John's, Fla	940
Washington, D. C	917

HUNTSVILLE, ALA., TO—

Jackson, Hinds, Miss	350
Key West, Monroe, Fla	1,109
Lebanon, Marion, Ky,	339
Louisville, Jefferson, Ky	332
Meridian, Lauderdale, Miss	254
Mobile, Mobile, Ala	392
Mount Sterling, Montgomery, Ky	457
Nashville, Davidson, Tenn	147
Paducah, McCracken, Ky	319
Savannah, Chatham, Ga	539
Shelbyville, Shelby, Ky	363
Saint Augustine, St. John's, Fla	747
Washington, D. C	724

INDIANOLA, TEX., TO—

	Miles.
Laredo, (Fort McIntosh,) Webb, Tex	293
Lavaca, Calhoun, Tex	16
Liberty, Liberty, Tex	291
Marlin, Falls, Tex	314
Marshall, Harrison, Tex	465
Navasota, Grimes, Tex	251
New Orleans, Orleans, La	432
Orange, Orange, Tex	291
Ringgold Barracks, Starr, Tex	274
Rio Grande City, Starr, Tex	273
San Antonio, Bexar, Tex	158
Victoria, Victoria, Tex	46
Waco, McLennan, Tex	344
Washington, D. C	1,628
Waxahachie, Ellis, Tex	369

JACKSBOROUGH, TEX., TO—

Dallas, Dallas, Tex	100
Denison, Tex	100
Laredo, (Fort McIntosh,) Webb, Tex	495
Longview, Upshur, Tex	257
Marlin, Falls, Tex	186
New Orleans, Orleans, La	700
Orange, Orange, Tex	457
Ringgold Barracks, Starr, Tex	636
Rio Grande City, Starr, Tex	635
San Antonio, Bexar, Tex	330
Sherman, Grayson, Tex	102
Vicksburgh, Warren, Miss	517
Victoria, Victoria, Tex	389
Waco, McLennan, Tex	166
Washington, D. C	1,591
Weatherford, Parker, Tex	40

JACKSON, MISS., TO—

Key West, Monroe, Fla	783
Lebanon, Marion, Ky	579
Louisville, Jefferson, Ky	572
Meridian, Lauderdale, Miss	96
Mobile, Mobile, Ala	323
Mount Sterling, Montgomery, Ky	700
Nashville, Davidson, Tenn	428
Paducah, McCracken, Ky	378
Saint Augustine, St. John's, Fla	857
Savannah, Chatham, Ga	641
Shelbyville, Shelby, Ky	603
Washington, D. C	1,016

JACKSON BARRACKS, LA., TO—

	Miles.
Baton Rouge, East Baton Rouge, La	133
Mount Vernon Arsenal, Mobile, Ala	188
New Orleans, Orleans, La	3
Washington, D. C	1,199

JACKONVILLE, OREG., TO—

Kelton, Box Elder, Utah	1,008
Lewiston, Nez Percés, Idaho	655
Marysville, Yuba, Cal	283
Monticello, Cowlitz, Wash	348
Ogden City, Weber, Utah	1,032
Olympia, Thurston, Wash	433
Omaha City, Douglas, Nebr	2,061
Portland, Multnomah, Oreg	295
Port Townsend, Jefferson, Wash	525
Red Bluff, Tehama, Cal	187
Reno, Washoe, Nev	443
Sacramento, Sacramento, Cal	325
Sacramento Junction, Placer, Cal	307
San Francisco, San Francisco, Cal	394
San Juan Island, Whatcom, Wash	555
Sitka, Alaska	1,083
Umatilla, Umatilla, Oreg	515
Vancouver, Clarke, Wash	313
Victoria, Brit. Col	575
Walla Walla, Walla Walla, Wash	570
Wallula, Walla Walla, Wash	540
Washington, D. C	3,330
Winnemucca, Humboldt, Nev	613

JEFFERSON BARRACKS, MO., TO—

Saint Louis, Saint Louis, Mo	12
Washington, D. C	918

KELTON, UTAH, TO—

Leavenworth City, Leavenworth, Kans	1,301
Ogden City, Weber, Utah	92
Omaha City, Douglas, Nebr	1,121
Portland, Multnomah, Oreg	750
Reno, Washoe, Nev	498
Sacramento, Sacramento, Cal	652
Sacramento Junction, Placer, Cal	634
San Francisco, San Francisco, Cal	735
Sioux City, Woodbury, Iowa	1,187
Washington, D. C	2,387
Winnemucca, Humboldt, Nev	328

KENNEBEC ARSENAL, ME., TO—

	Miles.
Augusta, Kennebec, Me	1

KIT CARSON, COLO., TO—

Laramie City, Albany, Wyo	309
Lawrance, Douglas, Kans	449
Leavenworth City, Leavenworth, Kans	482
Mesilla, Doña Aña, N. Mex	754
Omaha City, Douglas, Nebr	693
Pueblo, Pueblo, Colo	148
Salt Lake City, Salt Lake, Utah	803
San Francisco, San Francisco, Cal	1,592
Santa Fé, Santa Fé, N. Mex	444
Trinidad, Las Animas, Colo	168
Washington, D. C	1,678

LEBANON, KY., TO—

Louisville, Jefferson, Ky	67
Meridian, Lauderdale, Miss	570
Mobile, Mobile, Ala	683
Mount Sterling, Montgomery, Ky	139
Nashville, Davidson, Tenn	192
Paducah, McCracken, Ky	294
San Augustine, St. John's, Fla	992
Savannah, Chatham, Ga	784
Shelbyville, Shelby, Ky	98
Washington, D. C	787

LITTLE ROCK, ARK., TO—

Benton, Saline, Ark	26
Dardanelle, Yell, Ark	95
Memphis, Shelby, Tenn	135
New Orleans, Orleans, La	528
Shreveport, Caddo, La	240
Vicksburgh, Warren, Miss	390
Washington, Hempstead, Ark	130
Washington, D. C	1,072

LOS ANGELES, CAL., TO—

Bellena, San Diego, Cal	176
Cucamonga, San Bernardino, Cal	41
Guahonga, San Diego, Cal	137
Julian, San Diego, Cal	188
Maricopa Wells, Pima, Ariz	487

LOS ANGELES, CAL., TO—

	Miles.
Monte, Los Angeles, Cal	11
Oak Grove, San Diego, Cal	147
Prescott, Yavapai, Ariz	483
Rincon, San Bernardino, Cal	41
Riverside, San Bernardino, Cal	55
Sacramento, Sacramento, Cal	530
San Bernardino, Sen Bernardino, Cal	62
San Diego, San Diego, Cal	146
San Francisco, San Francisco, Cal	447
San Gabriel, Los Angeles, Cal	6
San Jacinto, San Diego, Cal	97
San Luis Obispo, San Luis Obispo, Cal	200
Santa Barbara, Santa Barbara, Cal	100
Spadra, Los Angeles, Cal	29
Temecula, San Diego, Cal	122
Tucson, Pima, Ariz	612
Warner's Ranch, San Diego, Cal	161
Washington, D. C	3,460
Wickenburgh, Yavapai, Ariz	392
Wilmington, Los Angeles, Cal	20

MADISON BARRACKS, SACKETT'S HARBOR, N. Y., TO—

Adams, Jefferson, N. Y	15
Buffalo, Erie, N. Y	247
De Kalb Junction, St. Lawrence, N. Y	61
Detroit, Wayne, Mich	485
Ogdensburgh, St. Lawrence, N. Y	80
Oswego, Oswego, N. Y	62
Plattsburgh Barracks, Clinton, N. Y	188
Potsdam, St. Lawrence, N. Y	78
Potsdam Junction, St. Lawrence, N. Y	85
Richland, Oswego, N. Y	33
Rome, Oneida, N. Y	74
Washington, D. C	548
Watertown, Jefferson, N. Y	10½

MARICOPA WELLS, ARIZ., TO—

Ogden City, Weber, Utah	980
Prescott, Yavapai, Ariz	186
San Bernardino, San Bernardino, Cal	425
San Diego, San Diego, Cal	397
San Francisco, San Francisco, Cal	934
Tucson, Pima, Ariz	125
Washington, D. C	3,278
Wickenburgh, Yavapai, Ariz	95
Wilmington, Los Angeles, Cal	507

MARSHALL, TEX., TO—

	Miles.
Jefferson, Marion, Tex	16
Navasota, Grimes, Tex	214
New Orleans, Orleans, La	464
Ringgold Barracks, Starr, Tex	737
Rio Grande City, Starr, Tex	736
San Antonio, Bexar, Tex	502
Sherman, Grayson, Tex	204
Shreveport, Caddo, La	42
Tyler, Smith, Tex	65
Vicksburgh, Warren, Miss	236
Victoria, Victoria, Tex	458
Waco, McLennan, Tex	210
Washington, D. C	1,297

MARYSVILLE, CAL., TO—

Davis Junction, Yolo, Cal	42
Eugene City, Lane, Oreg	399
Monticello, Napa, Cal	631
Ogden City, Weber, Utah	759
Olympia, Thurston, Wash	776
Omaha City, Douglas, Nebr	1,788
Portland, Multnomah, Oreg	578
Port Townsend, Jefferson, Wash	726
Reno, Washoe, Nev	170
Red Bluff, Tehama, Cal	86
Sacramento, Sacramento, Cal	52
San Francisco, San Francisco, Cal	111
San Juan Island, Whatcom, Wash	838
Sitka, Alaska	1,366
Umatilla, Umatilla, Oreg	798
Vancouver, Clarke, Wash	596
Victoria, Brit. Col	858
Walla Walla, Walla Walla, Wash	853
Wallula, Walla Walla, Wash	833
Washington, D. C	3,057
Winnemucca, Humboldt, Nev	340

MOUNT STERLING, KY., TO—

Nashville, Davidson, Tenn	313
Paducah, McCracken, Ky	369
Savannah, Chatham, Ga	905
Shelbyville, Shelby, Ky	81
San Augustine, St. John's, Fla	1,113
Washington, D. C	747

MOUNT VERNON ARSENAL, ALA., TO—

Mobile, Mobile, Ala	
Montgomery, Montgomery, Ala	
New Orleans, Orleans, La	
Washington, D. C	

NAVASOTA, TEX., TO—

	Miles.
Orange, Orange, Tex	180
Ringgold Barracks, Starr, Tex	525
Rio Grande City, Starr, Tex	524
San Antonio, Bexar, Tex	211
Sherman, Grayson, Tex	273
Shreveport, Caddo, La	256
Vicksburgh, Warren, Miss	450
Victoria, Victoria, Tex	167
Waco, McLennan, Tex	121
Washington, D. C	1,619

NEWPORT BARRACKS, KY., TO—

Cincinnati, Hamilton, Ohio	1
Covington, Kenton, Ky	1
Louisville, Jefferson, Ky	108
Washington, D. C	565

OGDEN CITY, UTAH, TO—

Olympia, Thurston, Wash	980
Omaha City, Douglas, Nebr	1,029
Portland, Multnomah, Oreg	843
Reno, Washoe, Nev	589
Sacramento, Sacramento, Cal	743
Sacramento Junction, Placer, Cal	725
San Francisco, San Francisco, Cal	826
San Juan Island, Whatcom, Wash	1,054
Santa Fé, Santa Fé, N. Mex	1,108
Sitka, Alaska	2,082
Sioux City, Woodbury, Iowa	1,095
Susanville, Lassen, Cal	687
Umatilla, Umatilla, Oreg	668
Vancouver, Clarke, Wash	774
Victoria, Brit. Col	1,074
Walla Walla, Walla Walla, Wash	613
Washington, D. C	2,298
Winnemucca, Humboldt, Nev	420

OLYMPIA, WASH., TO—

Portland, Multnomah, Oreg	138
Port Townsend, Jefferson, Wash	92
Red Bluff, Tehama, Cal	620
Reno, Washoe, Nev	876
Sacramento, Sacramento, Cal	758
Sacramento Juction, Placer, Cal	740
San Francisco, San Francisco, Cal	817
San Juan Island, Whatcom, Wash	122
Sitka, Alaska	1,150

OLYMPIA, WASH., TO—

	Miles.
Umatilla, Umatilla, Oreg	358
Vancouver, Clarke, Wash	156
Victoria, Brit. Col	142
Walla Walla, Walla Walla, Wash	413
Wallula, Walla Walla, Wash	383
Washington, D. C	3,278
Winnemucca, Humboldt, Nev	918

PADUCAH, KY., TO—

Savannah, Chatham, Ga	766
Shelbyville, Shelby, Ky	258
Washington, D. C	947

PIKESVILLE ARSENAL, MD., TO—

Baltimore, Baltimore, Md	8
Boston, Suffolk, Mass	434
Philadelphio, Philadelphia, Pa	109
Plattsburgh Barracks, Clinton, N. Y	511
Raleigh, Wake, N. C	362
Rutherfordton, Rutherford, N. C	540
Washington, D. C	46
Watertown Arsenal, Middlesex, Mass	442
Willett's Point, Queens, N. Y	218
Wilmington, New Hanover, N. C	427

PLATTSBURGH BARRACKS, N. Y., TO—

Raleigh, Wake, N. C	
Rutherfordton, Rutherford, N. C	
Washington, D. C	
Watertown Arsenal, Middlesex, Mass	
Willett's Point, Queens, N. Y	
Wilmington, New Hanover, N. C	

PORTLAND, OREG., TO—

Port Townsend, Jefferson, Wash	230
Red Bluff, Tehama, Cal	482
Reno, Washoe, Nev	738
Sacramento, Sacramento, Cal	620
Sacramento Junction, Placer, Cal	602
San Francisco, San Francisco, Cal	679
San Juan Island, Whatcom, Wash	260
Sitka, Alaska	1,288
Umatilla, Umatilla, Oreg	220
Vancouver, Clarke, Wash	18
Victoria, Brit. Col	280
Walla Walla, Walla Walla, Wash	275
Wallula Walla Walla, Wash	245
Washington, D. C	3,137
Winnemucca, Humboldt, Nev	780

PRESCOTT, (FORT WHIPPLE,) ARIZ., TO—

	Miles.
Sacramento, Sacramento, Cal	1,013
San Bernardino, San Bernardino, Cal	421
San Diego, San Diego, Cal	515
San Francisco, San Francisco, Cal	930
Tucson, Pima, Ariz	311
Washington, D. C	3,155
Wickenburgh, Yavapai, Ariz	91
Wilmington, Los Angeles, Cal	503

RED WILLOW, NEBR., TO—

Fort McPherson, Lincoln, Nebr	60
Omaha City, Douglas, Nebr	343
Washington, D. C	1,609

RENO, NEV., TO—

Sacramento, Sacramento, Cal	154
Sacramento Junction, Placer, Cal	136
San Francisco, San Francisco, Cal	237
San Juan Island, Whatcom, Wash	008
Sitka, Alaska	2,026
Susanville, Lassen, Cal	98
Umatilla, Umatilla, Oreg	776
Vancouver, Clarke, Wash	756
Victoria, Brit. Col	1,018
Walla Walla, Walla Walla, Wash	721
Washington, D. C	2,887
Winnemucca, Humboldt, Nev	170

RINGGOLD BARRACKS, TEX., TO—

Rio Grande City, Starr, Tex	1
San Antonio, Bexar, Tex	306
Sherman, Grayson, Tex	640
Shreveport, Caddo, La	731
Tyler, Smith, Tex	625
Victoria, Victoria, Tex	320
Waco, McLennan, Tex	480
Washington, D. C	1,902
Weatherford, Parker, Tex	597

RIO GRANDE CITY, TEX., TO—

	Miles.
San Antonio, Bexar, Tex	305
Sherman, Grayson, Tex	645
Shreveport, Caddo, La	730
Tyler, Smith, Tex	624
Victoria, Victoria, Tex	319
Waco, McLennan, Tex	479
Washington, D. C	1,901
Weatherford, Parker, Tex	596

RUTHERFORDTON, N. C., TO—

Baltimore, Baltimore, Md	532
Boston, Suffolk, Mass	958
Charlotte, Mecklenburgh, N. C	80
Cherryville, Gaston, N. C	37
Rockingham, Richmond, N. C	159
Wadesborough, Anson, N. C	138
Washington, D. C	478
Wilmington, New Hanover, N. C	276

SACRAMENTO, CAL., TO—

Sacramento Junction, Placer, Cal	18
San Andreas, Calaveras, Cal	70
San Bernardino, San Bernardino, Cal	592
San Diego, San Diego, Cal	600
San Francisco, San Francisco, Cal	83
Silver City, Lyon, Nev	534
Sitka, Alaska	1,640
Spokan Bridge, Stevens, Wash	1,050
Stockton, San Joaquin, Cal	47
Tucson, Pima, Ariz	1,142
Vallejo, Solano, Cal	60
Walla Walla, Walla Walla, Wash	895
Wallula, Walla Walla, Wash	865
Washington, D. C	3,041
Wickenburgh, Yavapai, Ariz	922
Wilmington, Los Angeles, Cal	510
Winnemucca, Humboldt, Nev	324

SAN ANTONIO, TEX., TO—

Austin, Travis, Tex	77
Belton, Bell, Tex	135
Boerne, Kendall, Tex	29
Bonham, Fannin, Tex	348
Boston, Bowie, Tex	433
Brazos Santiago, Cameron, Tex	355
Bremond, Robertson, Tex	189
Brenham, Washington, Tex	166

SAN ANTONIO, TEX., TO—

	Miles.
Brownsville, Cameron, Tex	322
Cherry Spring, Gillespie, Tex	91
Clarksville, Red River, Tex	392
Comfort, Kerr, Tex	45
Columbia, Brazoria, Tex	260
Columbus, Colorado, Tex	140
Corpus Christi, Nueces, Tex	162
Corsicana, Navarro, Tex	234
El Paso, El Paso, Tex	704
Fort Arbuckle, Chickasaw N., Ind. Ter	545
Fort Bliss, El Paso, Tex	701
Fort Clark, Kinney, Tex	126
Fort Concho, Bexar, Tex	229
Fort Davis, Presidio, Tex	479
Fort Duncan, (Eagle Pass,) Maverick, Tex	171
Fort Gibson, Cherokee N., Ind. Ter	489
Fort Griffin, Shackleford, Tex	369
Fort McIntosh, Webb, Tex	165
Fort McKavett, Menard, Tex	175
Fort Quitman, El Paso, Tex	619
Fort Richardson, Jack, Tex	331
Fort Smith, Sebastian, Ark	533
Fort Stockton, Pecos, Tex	399
Fredericksburgh, Gillespie, Tex	71
Galveston, Galveston, Tex	273
Hallsville, Harrison, Tex	369
Harrisburgh, Harris, Tex	224
Hillsborough, Hill, Tex	210
Houston, Harris, Tex	224
Indianola, Calhoun, Tex	158
Jacksborough, Jack, Tex	330
Leon Springs, Bexar, Tex	502
Marshall, Harrison, Tex	502
Mason, Mason, Tex	115
Menardsville, Menard, Tex	153
Navasota, Grimes, Tex	211
Ringgold Barracks, Starr, Tex	306
Rio Grande City, Starr, Tex	305
San Bernardino, San Bernardino, Cal	1,602
Sherman, Grayson, Tex	350
Shreveport, Caddo, La	425
Sulphur Springs, Hopkins, Tex	337
Victorio, Victoria, Tex	112
Waco, McLennan, Tex	174
Washington, D. C	1,771

SAN BERNARDINO, CAL., TO—

Bellena, San Diego, Cal	114
Cucamonga, San Bernardino, Cal	22
Guahonga, San Diego, Cal	75
Juham, San Diego, Cal	126
Maricopa Wells, Pima, Ariz	425
Monte, Los Angeles, Cal	52
Oak Grove, San Diega, Cal	85
Prescott, Yavapai, Ariz	421

SAN BERNARDINO, CAL., TO—

	Miles.
Rincon, San Bernardino, Cal	26
Riverside, San Bernardino, Cal	12
Sacramento, Sacramento, Cal	592
San Francisco, San Francisco, Cal	509
San Gabriel, Los Angeles, Cal	57
San Jacinto, San Diego, Cal	35
San Luis Obispo, San Luis Obispo, Cal	262
Santa Barbara, Santa Barbara, Cal	162
Spradra, Los Angeles, Cal	38
Temecula, San Diego, Cal	60
Tucson, Pima, Ariz	550
Warner's Ranch, San Diego, Cal	99
Washington, D. C	3,522
Wickenburgh, Yavapai, Ariz	330
Wilmington, Los Angeles, Cal	82

SAN DIEGO, CAL., TO—

Bellena, San Diego, Cal	64
Cucamonga, San Bernardino, Cal	187
Guahonga, San Diego, Cal	103
Juham, San Diego, Cal	52
Maricopa Wells, Pima, Ariz	397
Monte, Los Angeles, Cal	157
Oak Grove, San Diego, Cal	93
Prescott, Yavapai, Ariz	515
Rincon, San Bernardino, Cal	187
Riverside, San Bernardino, Cal	190
Sacramento, Sacramento, Cal	600
San Bernardino, San Bernardino, Cal	178
San Francisco, San Francisco, Cal	517
San Gabriel, Los Angeles, Cal	152
San Jacinto, San Diego, Cal	143
San Luis Obispo, San Luis Obispo, Cal	334
Santa Barbara, Santa Barbara, Cal	234
Spadra, Los Angeles, Cal	175
Temecula, San Diego, Cal	118
Tucson, Pima, Ariz	522
Washington, D. C	3,641
Wickenburgh, Yavapai, Ariz	444
Wilmington, Los Angeles, Cal	166

SAN FRANCISCO, CAL., TO—

Alameda, Alameda, Cal	11
Alcatras Island, San Francisco, Cal	3
Astoria, Clatsop, Oreg	642
Angel Island, San Francisco, Cal	6
Big Trees, Calaveras, Cal	117
Boisé City, Ada, Idaho	687
Calistoga, Napa, Cal	66
Camp Apache, Maricopa, Ariz	1,233
Camp Bidwell, Siskiyou, Cal	395

SAN FRANCISCO, CAL., TO—

	Miles.
Camp Bowie, Prince, Ariz	1,151
Camp Brown, ———, Wyo	1,139
Camp Crittenden, Pima, Ariz	1,090
Camp Date Creek, Yavapai, Ariz	866
Camp Douglas, Salt Lake, Utah	865
Camp Gaston, Klamath, Cal	334
Camp Grant, Pima, Ariz	1,005
Camp Halleck, Elko, Nev	581
Camp Harney, Grant, Oreg	672
Camp Haulpai, Yavapai, Ariz	885
Camp Independence, Inyo, Cal	524
Camp Lowell, (Tucson,) Pima, Ariz	1,039
Camp McDermitt, Humboldt, Nev	487
Camp McDowell, Maricopa, Ariz	948
Camp Mohave, Mohave, Ariz	759
Camp Stambaugh, Sweetwater, Wyo	1,107
Camp Three Forks, Owyhee, Idaho	652
Camp Verde, Yavapai, Ariz	959
Camp Warner, Grant, Oreg	540
Camp Wright, Mendocino, Cal	195
Canyon City, Grant, Oreg	887
Cape Disappointment, Pacific, Wash	657
Cheyenne, Laramie, Wyo	1,337
Chualar, Monterey, Cal	118
Cloverdale, Sonoma, Cal	101
Corinne, Box Elder, Utah	802
Cucamonga, San Bernardino, Cal	488
Denver, Arapaho, Colo	1,441
Drum Barracks, Los Angeles, Cal	428
Elko, Elko, Nev	531
Eugene City, Lane, Oreg	563
Fort Boisé, Ada, Idaho	687
Fort Bridger, Uintah, Wyo	964
Fort Colville, Stevens, Wash	1,207
Fort D. A. Russell, Laramie, Wyo	1,341
Fort Fred Steele, Carbon, Wyo	1,161
Fort Hall, Oneida, Idaho	942
Fort Klamath, Wasco, Oreg	484
Fort Lapwai, Nez Percés, Idaho	1,051
Fort Point, San Francisco, Cal	7
Fort Steilacoom, Pierce, Wash	809
Fort Stevens, Clatsop, Oreg	650
Fort Tongass, Alaska	1,535
Fort Whipple, Yavapai, Ariz	930
Fort Wrangel, Alaska	1,713
Fort Yuma, San Diego, Cal	714
Fremont, Dodge, Nebr	1,808
Gabilan, Monterey, Cal	111
Galt, Sacramento, Cal	112
Gilroy, Santa Clara, Cal	80
Healdsburgh, Sonoma, Cal	83
Hollister, Monterey, Cal	96
Jacksonville, Jackson, Oreg	384
Kansas City, Jackson, Mo	2,059
Kelton, Box Elder, Utah	735
Lake Tahoe, Placer, Cal	217
La Paz, Yuma, Ariz	719
Los Angeles, Los Angeles, Cal	447
Lathrop, San Joaquin, Cal	81
Lime Point, San Francisco, Cal	8

SAN FRANCISCO, CAL., TO—

	Miles.
Marysville, Yuba, Cal	111
Maricopa Wells, Pima, Ariz	934
Missouri Valley Junction, Harrison, Iowa	1,846
Modesto, Stanislaus, Cal	101
Monte, Los Angeles, Cal	458
Monterey, Monterey, Cal	133
Napa, Napa, Cal	39
Napa Junction, Napa, Cal	30
Natividad, Monterey, Cal	104
New York City, New York, N. Y	3,250
Oakland, Alameda, Cal	8
Ogden City, Weber, Utah	826
Olympia, Thurston, Wash	817
Omaha City, Douglas, Nebr	1,855
Petaluma, Sonoma, Cal	51
Philadelphia, Philadelphia, Pa	3,175
Point St. José, San Francisco, Cal	3
Portland, Multnomah, Oreg	679
Prescott, Yavapai, Ariz	930
Presidio, San Francisco, Cal	5
Red Bluff, Tehama, Cal	197
Reno, Washoe, Nev	237
Riverside, San Bernardino, Cal	502
Sacramento, Sacramento, Cal	83
St. Louis, St. Louis, Mo	2,341
St. Paul, (Kodiak Island,) Alaska	2,225
San Bernardino, San Bernardino, Cal	509
San Diego, San Diego, Cal	517
San Gabriel, Los Angeles, Cal	453
San Jacinto, San Diego, Cal	544
San José, Santa Clara, Cal	47
San Juan, Monterey, Cal	92
San Juan Island, Whatcom, Wash	905
San Luis Obispo, San Luis Obispo, Cal	245
San Mateo, San Mateo, Cal	21
Santa Barbara, Santa Barbara, Cal	345
Santa Cruz, Santa Cruz, Cal	82
Sioux City, Woodbury, Iowa	1,921
Sitka, Alaska	1,557
Spadra, Los Angeles, Cal	476
Stockton, San Joaquin, Cal	47
Susanville, Lassen, Cal	240
Temecula, San Diego, Cal	604
The Dalles, Wasco, Oreg	799
Truckee, Washoe, Nev	203
Tucson, Pima, Ariz	1,039
Umatilla, Umatilla, Oreg	899
Valejo, Solano, Cal	23
Vancouver, Clarke, Wash	697
Walla Walla, Walla Walla, Wash	954
Wallula, Walla Walla, Wash	924
Washington, D. C	3,126
Wickenburgh, Yavapai, Ariz	839
Wilmington, Los Angeles, Cal	427
Winnemucca, Humboldt, Nev	407
Yerba Buena Island, San Francisco, Cal	3
Yosemite, Mariposa, Cal	199

SAN JUAN ISLAND, WASH., TO—

	Miles.
Sitka, Alaska	1,028
Umatilla, Umatilla, Oreg	480
Vancouver, Clarke, Wash	278
Victoria, Brit. Col	20
Walla Walla, Walla Walla, Wash	535
Wallula, Walla Walla, Wash	505
Washington, D. C	3,400
Winnemucca, Humboldt, Nev	1,040

SANTA FÉ, N. MEX., TO—

Albuquerque, Bernalillo, N. Mex	71
Bent's Fort, Bent, Colo	374
Booneville, Pueblo, Colo	393
Cimarron, Mora, N. Mex	208
Clifton, Colfax, N. Mex	247
Colorado City, El Paso, Colo	417
Denver, Arapahoe, Colo	493
El Paso, El Paso, Tex	360
Fort Bascom, San Miguel, N. Mex	251
Fort Bayard, Grant, N. Mex	415
Fort Bliss, El Paso, Tex	363
Fort Bowie, Pima, Ariz	498
Fort Craig, Socorro, N. Mex	189
Fort Cummings, Grant, N. Mex	370
Fort Garland, Costilla, Colo	163
Fort Leavenworth, Leavenworth, Kans	937
Fort Lyon, Bent, Colo	392
Fort McRae, Socorro, N. Mex	221
Fort Reynolds, Pueblo, Colo	391
Fort Selden, Doña Aña, N. Mex	289
Fort Stanton, Socorro, N. Mex	239
Fort Sumner, San Miguel, N. Mex	180
Fort Union, Mora, N. Mex	106
Fort Wingate, Santa Aña, N. Mex	183
Greenhorn, Huerfano, Colo	343
Kansas City, Jackson, Mo	931
Kit Carson, Greenwood, Colo	442
Las Cruces, Doña Aña, N. Mex	307
La Junta, Mora, N. Mex	97
Leesburgh, (Fort Selden,) Doña Aña, N. Mex	290

SITKA, ALASKA, TO—

Fort Kenay, (Nicholas,) Alaska	784
Oonalaska, Alaska	1,268
Saint George Island, Alaska	1,449
Saint Paul Island, Alaska	1,448
Saint Paul, (Kodiak Island, Fort Kodiak,) Alaska	668
San Francisco, San Francisco, Cal	1,557
Umatilla, Umatilla, Wash	1,468
Vancouver, Clarke, Wash	1,306
Walla Walla, Walla Walla, Wash	1,563
Wallula, Walla Walla, Wash	1,533
Washington, D. C	4,212
Winnemucca, Humboldt, Nev	2,068

SIOUX CITY, IOWA, TO—

	Miles.
Sioux Falls, Minnehaha, Dak	84
Swan Lake, Lincoln, Dak	14
Vermillion, Clay, Dak	27
Washington, D. C	1,314
Whetstone Agency, Gregory, Dak	151
Yankton, Yankton, Dak	64

SHREVEPORT, LA., TO—

Albany, Albany, N. Y	1,649
Alexandria, Rapides, La	198
Chicago, Cook, Ill	991
Herne, Robertson, Tex	241
Little Rock, Pulaski, Ark	238
Longview, Upshur, Tex	66
Marshall, Harrison, Tex	42
Monroe, Ouachita, La	120
New Orleans, Orleans, La	443
Palestine, Anderson, Tex	150
Vicksburgh, Warren, Miss	194
Washington, D. C	1,255

SHERMAN, TEX., TO—

Shreveport, Caddo, La	246
Victoria, Victoria, Tex	399
Waco, McLennan, Tex	266
Washington, D. C	1,479

THE DALLES, OREG., TO—

Elko, Elko, Nev	804
Eugene City, Lane, Oreg	236
Fort Boisé, Ada, Idaho	380
Fort Colville, Stevens, Wash	408
Fort Hall, Oneida, Idaho	837
Fort Klamath, Wasco, Oreg	515
Fort Lapwai, Nez Percés, Idaho	250
Fort Leavenworth, Leavenworth, Kans	1,934
Fort Stevens, Clatsop, Oreg	224
Fort Tongass, Alaska	1,000
Fort Wrangel, Alaska	1,178
Jacksonville, Jackson, Oreg	415
Kelton, Box Elder, Utah	630
Lewiston, Nez Percés, Idaho	240
Marysville, Yuba, Cal	688
Monticello, Cowlitz, Wash	173
Olympia, Thurston, Wash	258
Port Townsend, Jefferson, Wash	350
Portland, Multnomah, Oreg	120
Red Bluff, Tehama, Cal	602
Sacramento, Sacramento, Cal	740
Sacramento Junction, Placer, Cal	722
San Francisco, San Francisco, Cal	799

THE DALLES, OREG., TO—

	Miles.
San Juan Island, Whatcom, Wash	370
Sitka, Alaska	1,408
Spokan Bridge, Stevens, Wash	310
Umatilla, Umatilla, Oreg	100
Vancouver, Clarke, Wash	192
Victoria, Brit. Col	370
Walla Walla, Walla Walla, Wash	155
Wallula, Walla Walla, Wash	125
Washington, D. C	3,020
Winnemucca, Humboldt, Nev	660

TUCSON, ARIZ., TO—

Fort Bliss, El Paso, Tex	353
Fort Davis, Presidio, Tex	573
Fort Yuma, San Diego, Cal	325
Guaymas, Mexico	351
La Libertad, Mexico	227
Lobos, Mexico	213
Maricopa Wells, Pima, Ariz	125
Mesilla, Doña Aña, N. Mex	300
Prescott, Yavapai, Ariz	311
San Antonio, Bexar, Tex	1,052
San Bernardino, San Bernardino, Cal	550
San Diego, San Diego, Cal	522
San Francisco, San Francisco, Cal	1,039
Santa Fé, Santa Fé, N. Mex	610
Washington, D. C	2,728
Wickenburgh, Yavapai, Ariz	220
Wilmington, Los Angeles, Cal	632

UMATILLA, OREG., TO—

Vancouver, Clarke, Wash	202
Victoria, Brit. Col	500
Walla Walla, Walla Walla, Wash	55
Wallula, Walla Walla, Wash	25
Washington, D. C	2,966
Winnemucca, Humboldt, Nev	606

VANCOUVER, WASH., TO—

Victoria, Brit. Col	298
Walla Walla, Walla Walla, Wash	257
Wallula, Walla Walla, Wash	227
Washington, D. C	3,122
Winnemucca, Humboldt, Nev	798

VICTORIA, TEX., TO—

Waco, McLennan, Tex	233
Washington, D. C	1,674

WACO, TEX., TO—

	Miles.
Washington, D. C	1,446

WALLA WALLA, WASH., TO—

Wallula, Walla Walla, Wash	30
Washington, D. C	2,911
Winnemucca, Humboldt, Nev	551

WALLULA, WASH., TO—

Washington, D. C	2,941
Winnemucca, Humboldt, Nev	581

WATERTOWN ARSENAL, MASS., TO—

Baltimore, Baltimore, Md	430
Boston, Suffolk, Mass	8
Washington, D. C	472
Willett's Point, Queens, N. Y	264
Wilmington, New Hanover, N. C	853

WATERVLIET ARSENAL, N. Y., TO—

Troy, Rensselaer, N. Y	1

WICKENBURGH, ARIZ., TO—

Washington, D. C	3,183
Wilmington, Los Angeles, Cal	412

WILLETT'S POINT, N. Y., TO—

Baltimore, Baltimore, Md	210
Boston, Suffolk, Mass	256
Brooklyn, Kings, N. Y	18
Flushing, Queens, N. Y	9
Flushing Junction, N. Y	8
New York, New York, N. Y	20
Whitestone, Queens, N. Y	2
Wilmington, New Hanover, N. C	629

WILMINGTON, CAL., TO—

Drum Barracks, Los Angeles, Cal	1
Washington, D. C	3,480

⊙

www.ingramcontent.com/pod-product-compliance
Lightning Source LLC
LaVergne TN
LVHW021410110826
845150LV00007B/1857

* 9 7 8 1 4 2 5 5 1 0 9 6 1 *